Agrarian Crisis:
Farmers' Suicides in Warangal District

By

Emmadi Naveen Kumar

CAMBRIDGE
SCHOLARS
PUBLISHING

Agrarian Crisis:
Farmers' Suicides in Warangal District,
by Emmadi Naveen Kumar

This book first published 2012

Cambridge Scholars Publishing

12 Back Chapman Street, Newcastle upon Tyne, NE6 2XX, UK

British Library Cataloguing in Publication Data
A catalogue record for this book is available from the British Library

ISBN (10): 1-4438-4040-8, ISBN (13): 978-1-4438-4040-8

Dedicated to my Brother
Emmadi Pavan Kumar

TABLE OF CONTENTS

ACKNOWLEDGEMENTS

"Gratitude is the language of the heart. When the language is from the heart, emotions prevail and when emotions prevail, words fail".

I extend my deep love and affection to my parents, my father Mr. Emmadi Das and my mother Mrs. Emmadi Sumithra, for giving me this wonderful life and for supporting me in every single second of it.

I am grateful to Cambridge Scholars Publication (CSP) for their significant suggestions that have helped augment the readability of this book and have brought in an unexpected dimension of support.

I express my respect, special regards and heartfelt thanks to Dr. Laxminarayana and Dr. R.V. Ramanamurthy, faculty members, Department of Economics, University of Hyderabad, for extending their valuable time, assistance, unfailing interest and constant encouragement throughout my work. I am extremely grateful to Dr. E. Revathi, faculty member at Centre for Economics and Social Studies (CESS) for the help she provided to me during my work. I am also grateful to Dr. A. Sunitha (Anveshi, Hyderabad), who took pains to go through my chapters and helped me put my ideas together in a consolidated manner.

I express my thanks to the staff of Chief Planning Officer, Warangal, District Revenue Office, Warangal; Directorate of Economics and Statistics, Andhra Pradesh; and Centre for Economics and Social Studies for extending their help during my research work.

I thank my entire family members, friends, my Department scholars, and my juniors (Mr. Ch. Sudhakar, Mrs. Srilatha, Mr. E. Srinivas, Mr. E. Satish, Mr. E. Sravan, Mrs. Elisha, Mrs. Sumalatha, Praneeth, Karthika, Teja, Mary Kezia, Anirudh Paul, Honey, Vamshi Krishna, Rachit Badolia, Gopal Agarwal, Rajinikanth, Harinath, G. Laxminarayana, Indrapal Yadav, Julie, Qurratulain Khan, Damerilin, Kerri, Queenie and Dr. Grace),who have extended their love and affection in my life.

Last but not the least, I thank my Department office staff and all other administration staff for providing me with their support in all my academic needs.

Emmadi Naveen Kumar

CHAPTER ONE

INTRODUCTION

1.1 Agriculture in India

India is primarily an agrarian country, and its economy is agrarian. Even now, 60% of India's population is dependent on agriculture for their livelihood. Since the time of the British Raj, in most parts of India, agriculture was an occupation of the small and middle farmers. But today the status of agriculture in India is dismal. Lack of funds to invest in the irrigation sector, mismanagement of land and constant droughts have brought about this dismal condition. Since the early nineties, the glory days of India's green revolution have diminished. Profits from the green revolution areas have reached a point of saturation, and this was mainly triggered by the decline in land and factor productivity from the late eighties. The production rate or yield rate of crops has been reported to be stagnant in recent years with no hope of increasing the land area under cultivation.

Liberalization and globalization have had a complex effect on the Indian agricultural sector and this is quite different from other nations who carried out radical land reforms before embarking upon macro economic reforms. The main causes for such a negative effect can be attributed to India's failure to make better policies, investment measures and domestic preparations before liberalizing the economy. Various planning policies have helped the nation achieve the stage that it has now reached, but the time has come for the role of the state to be redefined. The state should be given the role of a promoter cum facilitator of development. The so called market friendly policies that were designed to facilitate and diversify more private investment to make farming more viable and to compact land holdings will have to be revised.

Today, statistics show that the contribution made by the agricultural sector to the overall GDP has declined and the service sector has emerged as the primary contributor to GDP growth. This clearly indicates that income generation is not occurring in the agriculture sector or is taking place at a dismal rate. Growth in the agriculture sector is the only route to reducing poverty and to fulfilling the various social objectives plaguing India. The question that now lies before India is how best to frame and exercise the right policies and programmes to achieve the set goals.

1.2 The Role of Agriculture in Development

The role of agriculture in India's as well as Andhra Pradesh's economy is vital. The role that agriculture plays in the development of any economy will be clarified further in the discussion that follows. The contribution of agriculture to economic growth can be classified under the four following sections –

- Product Contribution
- Factor Contribution
- Market Contribution
- Foreign Exchange Contribution

For the development of other sectors of the economy, surplus agricultural production is extremely important. It has to be able to feed laborers working in alternative occupations. An excessive dependence on imports will only take a toll on the economy. Therefore, food production above subsistence level becomes extremely crucial as this alone will help enhance export activities and pay for foreign exchange. Marketable surplus is the main factor for economic growth as only this will trigger labour productivity. In case the agriculture sector fails to enhance production, compulsory taxation could be an option to support the economy. Lesser rate of production means a greater reduction in income generation and an increase in industrial wages. However, the most obvious result will be the increase of market prices which will directly affect the industry. The role played by agriculture in an economy's development can be seen in the case of England where the agricultural revolution helped develop the country's economy during the initial stages.

There are three aspects of factor contribution – labour and capital contribution. Here, labour contribution has been explained in terms of the Lewis model of development. The agricultural sector has to support labour for the industrial and its allied sectors. But for this, agricultural production should increase. Industrial expansion and development can take place with the cheap labour provided by the agricultural sector. It is also a good source of savings and capital accumulation for industrial development. The pattern of investments differs from one agriculturalist to another. The more prosperous agriculturalists invest in the industrial sector but the smaller peasants make deposits in rural banks. Government taxations on the agricultural sector can act as involuntary savings which can be used for developments in the industrial and agricultural sector. Earlier, due to the backward bending supply curve of effort, the government-supported prices were fixed at the lower rate than the market price for agricultural commodities. However, the statistics now show that farmers are affected by the price incentives. They change crops and increase or decrease the supply depending on the relative changes in prices.

The contribution of the agricultural sector to the economy is through their demand for industrial goods. The growth of the agricultural and the industrial sector is thus complementary. In other words, one depends on the other. It is often said that to increase the rate of industrialization, the growth and development of agricultural sectors is essential. Moreover, price policy has a great role to play here. If the prices of farm products are low from the supply side, it naturally boosts industrial growth as it makes the raw materials and goods cheaper. However, if it is low from the demand side it will only help reduce the purchasing power, and thereby lower the demand for industrial goods. Therefore, it is extremely necessary to make both sides of the agricultural sectors achieve a well balanced growth and development.

For any economy, agriculture is the primary source of foreign exchange. Earlier, the sources of foreign exchange were the primary products which helped us procure other commodities which we were unable to produce domestically. There is no doubt that a healthy inter-sector relationship is crucial for the development of an economy. For example, in order to divert the labour force from the agricultural sector to the non-agricultural sector, development levels of both sectors is extremely important. Only growth in the industrial sector can pull labour from the agricultural sector and for the industrial sector to grow, growth in the agricultural sector is necessary. Only growth in the agricultural sector can help in the creation and growth of the market. Therefore, a particular sector's role in an economy can only be judged by looking at its contribution to the growth of this economy. Thus, the role of agriculture should be judged on the basis of its contribution to helping the non-farm sectors grow.

1.3. Agrarian Crisis in India

> "Indian agriculture is heading for a crisis as food output stagnates and millions of poor farmers struggle with high debt and crop failures, Economic growth averaging 9% a year, fuelled by manufacturing and services, has masked the crisis in the countryside, "Green Revolution" of the 1960s that made India self-sufficient in food. While the contribution of manufacturing and services is laudable, it is still the farm sector that provides the largest employment in the subcontinent." (M.S. Swaminathan)

About two thirds of India's 1.1 billion people still depend on agriculture for their livelihood and are being neglected in the euphoria of a surging economy. But Indian agriculture, hostage to the vagaries of the monsoon, has been in decline in recent years and is growing at less than a quarter of the pace of the overall economy. Annual per capita food grain production declined from 207 kilograms (455 pounds) in 1995 to 186 kilograms in 2006. The rate of agricultural growth fell from 5% in the mid-1980s to less than 2% in the past half-decade. At the same time, thousands of debt-ridden farmers have committed suicide after crop failures forced many to sell plots they held for generations.

The people's protest against Special Economic Zones in various parts of the country, including at Nandigram in West Bengal, stagnation in agriculture, import of food grains, widespread suicide of farmers—all these are systems of simmering discontent in the agricultural sector. What is highlighted today in the national scene is the image of "incredible India" and "shining India". We hear often about India as a country with a very high economic growth, a country with the highest numbers of billionaires in Asia, and a country of world renowned information technology. But we do not hear enough about the serious problems in agriculture. Those who govern us do not seem to be concerned with this problem or they probably do not want to be. But we cannot easily ignore this problem any longer.

It was with the Structural Adjustment Programme (SAP) in 1991 that the policy of globalization was concretely introduced in India. Based on this policy and the directives of the World Bank, International Monetary Fund and World Trade Organization, the Indian economy was substantially overhauled. The Export-Import policy was liberalised, the import and customs duties of many products were drastically lowered or totally dropped so that they could be imported without any restriction. The government started reducing its investment in agriculture and the industrial sector, thereby allowing the private sector to take over. The restructuring of the public distribution system really affected the availability of food grains to the poor at subsidized rates. All such measures had implications for the farm sector.

Problems of the Agrarian Sector and their Consequences

Fifteen years of economic liberalization has adversely affected Indian agriculture. The most prominent manifestation of this is in the drastic decline in the growth rate of food grains. The rate of growth of agricultural output was gradually increasing in 1950-1990, and it was greater than the rate of growth of the population. In the 1980s the agricultural output grew at about four per cent per annum. Thus, India became self-sufficient with respect to food and started exporting wheat and rice. But during the 10-year period after the start of liberalization, the rate of growth declined to two per cent. According to the Mid-term Appraisal of the Tenth Five Year Plan (2002-07), the rate of growth of the GDP in agriculture and allied sectors was just one per cent per annum during the year 2002-05. As a result, per capita availability of food grains decreased, the growth rate of the population became higher than that of food grains and India started to import food grains at a much higher price than that of the domestic market.

Secondly, unemployment in the agricultural sector increased during the reform period. This was because agriculture was not profitable during this period due to the fall in the price of farm products. As a result, the number of people who were employed in the primary sector and the area under cultivation decreased, which in turn caused a decline in rural employment. According to the National Sample Survey, the annual rate of growth of employment in the rural areas was 2.07 per cent in 1987-1984, while it declined to a mere 0.66 per cent in 1993-2000, which corresponds to the period of liberalization. Thus, not only farmers but also Dalits and Tribes who were heavily dependent on agriculture, became unemployed.

Farmer suicide is the third fall-out of stagnation in agriculture. When agriculture was not yielding remunerative income, the life of farmers became very desperate. Many of them committed suicide as a last resort. As revealed by the Union Agricultural Minister in the Lok Sabha in 2004, over one lakh farmers committed suicide in India after the economic reform started. According to the National Crime Records Bureau, 17,060 farmers committed suicide in the country in 2006 with Maharashtra having the highest number of (4453) suicide deaths. Punjab and Andhra Pradesh also find a place in the list of States showing the highest numbers of farmer suicides. This marks a record in the agricultural history of India. It points to the acute nature of the problem that has affected the vast majority of the population, and which has created a real crisis. But unfortunately, the government and our country's people do not consider it a crisis. Their lack of seriousness and lukewarm response to the problem of farmer suicides is evidence of this reality.

In 2006, the government launched a multi-billion-dollar welfare drive promising 100 days of work for every rural family in the battle against rural poverty. But only about 3% of the households that signed up received employment for 100 days and many only received about two weeks of employment. This data comes from a six-month internal audit of the programme which cited widespread examples of corruption, inefficiency and misuse of funds for the programme's poor performance. The government has repeatedly said that it wants to make India's economic growth more inclusive. But the delivery system has collapsed because of "monstrous growth" in the bureaucracy. When the first Green Revolution was launched, it was carried out like a symphony in unison by scientists, policy makers, state agriculture departments, marketing agencies and farmers. While the government spearheads the establishment of Chinese-style special economic zones for industry, the time has come to set up special agricultural zones to boost farm production and reduce the widening income disparities between rural and urban India. The government should step in with administrative and infrastructural support and market access while scientists and agricultural think tanks contribute "technology and tools". (M.S. Swaminathan).

The distress of farmers in India can be traced back to the introduction of technology-led, capital intensive farming in the heyday of the Green Revolution. With the advent of 'economic liberalization' and the globalization of trade, this distress has been aggravated. Unfair rules of the multilateral global trading regime have depressed global and domestic prices, and have denied Indian farmers adequate remunerative prices. The poor farmer is squeezed between high input costs and low returns. Credit obtained from formal or informal banking systems is unable to bail him out of this precarious situation. Caught in a vicious debt trap, many farmers have resorted to suicide.

1.4. Reasons for the Agrarian Crisis in India

1.4.1 Demographic Pressure on Agriculture

During 1991-2000, 72 % of the population and 76% of the workforce in India were rural, accounting for about one-fifth of the national income. Even during the 1990s, the period of economic reforms and the much-lauded high economic growth, there was no substantial increase in the share of the rural non-farm sector. The employment status of rural labor tends towards relatively more insecure casual labor, while self-employment and regular employment indicate a declining share (Reddy & Srijit).

Table No: 1. Sectoral Share and Employment Status of Rural Workforce (Percent)

Rural employment	1983	1987-88	1993-94	1999-2000	2004-05
Agriculture	81.49	77.46	78.39	76.16	70.08
Non-agriculture	18.51	22.54	21.61	23.84	29.92
Status of rural workforce					
Self employed	61.37	59.50	57.96	55.76	60.2
Hired-regular	7.15	7.79	6.45	6.83	7.1
Hired-casual	31.49	32.72	35.59	37.41	32.8

Source: Bhalla (2005); NSSO, employment and Unemployment situation In India.

Table No: 2. Share of Agriculture in GDP and Employment (Percent)

Year	Share of Agriculture in GDP at 1999-2000 prices	Share of Agriculture in Employment (UPSS)
1972-73	41.0	73.9
1993-94	30.0	63.9
1999-2000	25.0	60.2
2004-05	20.2	56.5

Source: Central statistical Organization (CSO); National Accounts Statistics, various years; and NSSO, Employment and Unemployment Situation in India, various rounds.

An important structural feature of the Indian economy has been a continuous decline in the share of agriculture in total gross domestic product (GDP), but very slow diversification of the workforce away from agriculture. Agriculture's share in the GDP declined from 41% in 1972-73 to 20% in 2004-05.

1.4.2. Increasing Stress on Irrigation Resources

Unequal availability of irrigation across the country, and an increasing stress on available irrigation resources is one of the major reasons for the agrarian crisis in India. India is not very rich in irrigation resources. With 16% of the world's population, for instance, India is endowed with only 4% of the total available fresh water. Further more, the regional distribution of available water resources, including rainfall, is highly uneven. The recent trends in irrigation show the distortion present in the development and utilization of water resources for agricultural purposes. Dependence on groundwater has emerged as the single largest source of irrigation, with all its accompanying problems of serious risks to farmers' investments and degradation of the environment.

Table No: 3.

Year	Canals	Tanks	Wells and Tube wells	Others	Total	NIA/NSA (%)
1950-51	8,300 (39.71)	3,600 (17.22)	6,000 (28.71)	3,000 (14.35)	20,900 (100)	17.56
1970-71	12,838 (41.28	4,112 (13.22)	11,887 (38.22)	2,266 (7.29)	31,103	22.17
1980-81	15,292 (39.94)	3,182 (8.22)	17,695 (45.70)	2,551 (6.59)	38.720 (100)	27.66
190-91	17,453 (36.34)	2,944 (6.13)	24,694 (51.42)	2,932 (6.11)	48,023 (100)	33.41
2003-04	15,145 (27.48)	1,943 (3.53	35,265 (64.00)	2,752 (4.99)	55,105 (100)	40.06

Growth Rate (% per Annum)

1970s	1.76	-2.53	4.25	1.19	2.21	
1980s	1.33	-0.77	3.00	1.40	2.18	
1990-01 to 2003-04	-1.09	-3.15	2.80	-0.49	1.06	

Note: Canals include both government and private, but the latter's share is negligible. NIA denotes Net Irrigated Area, NSA denotes Net Sown Area. (Source: Ministry of Agriculture, Government of India)

1.4.3. Liberal Import of Agricultural Products

The main reason for the crash in prices of agricultural products in India, especially of cash crops, was the removal of all restrictions to import these products. As, for example, when the Government of India reduced the import duty on tea and coffee from Sri Lanka and Malaysia, their prices in the domestic market fell drastically. Thus, the cultivation of such products became unprofitable and so their production was fully or partly stopped. Since the removal of quantitative restrictions and lowering of import duties were in accordance with the restrictions of the World Trade Organization (WTO), the crash in the prices of agricultural products is directly related to the liberalization policy of the government.

1.4.4. Cutback in Agricultural Subsidies

In the post-reform period, the government reduced different types of subsidies to agriculture, and this has increased the production cost of cultivation. According to Ramesh Chand, an economist, cutback in subsidy and control of fertilizers over the last few years has adversely affected the agricultural sector. It has increased the input cost and has made agriculture less profitable. Since the decrease in subsidy to agriculture is part of the regulations of the WTO, it is related to the policy of globalization.

1.4.5. Lack of Formal Credit to Agriculture

After 1991, the lending pattern of commercial banks, including nationalized banks, to agriculture drastically changed and resulted in difficulties in obtaining loans and unaffordable interest rates. This has forced farmers to rely on moneylenders and has thus pushed up the expenditure on agriculture. The National Commission for Agriculture, headed by Dr M.S. Swaminathan, also pointed out that removal of lending facilities and concessions offered by banks during the post-reform period have accelerated the crisis in agriculture. When the farmers were not able to pay back loans with high interests, they fell into the debt trap. Studies show that most farmer suicides were due to the debt trap. It is part of the policy of privatization that banks, even nationalized banks, look for profit over their social responsibilities to the people.

1.4.6. Decline in Government Investment in the Agricultural Sector

Studies show that after the economic reforms started, the government's expenditure and investment in the agricultural sector have been drastically reduced. This is based on the policy of minimum intervention by the government enunciated by the policy of globalization. The expenditure of the government in rural development, including agriculture, irrigation, flood control, village industry, energy and transport, declined from an average of 14.5 per cent in 1986-1990 to six per cent in 1995-2000. When the economic reforms started, the annual rate of growth of irrigated land was 2.62 per cent. It later reduced to 0.5 per cent in the post-reform period. The consequences were many. The rate of capital formation in agriculture came down, and the agricultural growth rate was also adversely affected. This has affected the purchasing power of rural people and subsequently, their standard of living.

1.4.7. Special Economic Zones

As part of economic reforms, the system of land being taken over by the government for commercial and industrial purposes was introduced in the country. As per the Special Economic Zones Act of 2005, the government has so far notified about 400 such zones in the country. Very often it is fertile land which has been acquired. According to Khasanoki, the government has acquired five million hectares of land for purposes other than agriculture between 1991 and 2003. This is almost half of what was acquired during the last 40 years. It was reported in the news that the government decided to acquire 10,120 hectares of land near Mumbai (almost one-third the total area of Mumbai) for the Reliance Company and subsequently reduced it to 5000 hectors due to public pressure. Since the SEZ deprives farmers of their land and livelihood, it is harmful to agriculture. In order to promote export and industrial growth in line with globalization the SEZ was introduced in many countries.

1.5. Research Problem

There are considerable regional disparities in the agrarian economy of Andhra Pradesh. The early phase of green revolution in the state was largely confined to resource – rich South Coastal Andhra, and bypassed the semi-armed and rain fed areas of Rayalaseema, North and South Telangana, and North Costal Andhra. In the recent phase of the spread of the green revolution, practices took the form of farmers in dry regions adopting high value crops that would integrate them with markets that fetch higher returns from agriculture. Those farmers with a low resource base were forced to embark on high-risk investments, including those on groundwater development. The disparities in agriculture also took the form of disparities in vulnerability; the suicides of farmers in the state reflect these regional dimensions as well.

The first spurt in farmers' suicides occurred in the mid-1980s (1986-87) and there were a few cases in the later years. It is the second spurt, beginning with 1997 that became a recurring phenomenon, between 1998 and 2006 there were 4403 farmers' suicides in the state. According to the estimates of May 2004 and 10th November 2005 alone there were 1068 farmer suicides, and in addition there were 277 weaver starvation deaths in the same period (Kumar 2005). The commission also brings out the fact that the state was rocked by the revelation that 26 debt-ridden farmers of Guntur district had sold their kidneys.

There have been a number of studies in the state based on a sample investigation of households of the victims to analyse the proximate causes for the suicides. A few of them are reviewed here to understand the nature of the stress factors behind the suicides. In 1998, a people's tribunal (RSC 1998) heard the depositions from 60 farming households of victims drawn from across five districts in the state. Of the 60, the majority (42) reported their dependence on wells or bore well for irrigation, on which they invested a substantial part of their resources. Thirty-three of them cited water shortage as the main reason for their crop failure and that this triggered the suicides. Of the 21, who reported heavy borrowing for investing on the wells or bore wells or deepening of bore wells, only one reported bank credit, while the rest had borrowed from private informal sources at very high interest rates.

Even lack of proper institutional credit and proper public investment in the agriculture sector, which is meant to be the backbone of the Indian economy, is leading the agriculture sector into an un-healable condition. The agrarian economy in Andhra Pradesh, as in the rest of India is diverse and is built on regional variations. In additional to this, the green revolution in the state was confined to resource rich areas, and bypassed vast dry land zones mostly located in Telangana and Rayalaseema regions, thus increasing regional and social disparities. Further, in Andhra Pradesh, the forerunner of neo-liberal reforms vigorously implemented by the earlier government, important changes took place in agriculture in terms of the shrinking role of the government and the increased exposure of the farmer community to the markets. It was argued that these reforms pushed the agricultural sector and especially the dry land regions into crisis. The crisis manifested itself in an extreme form in suicides by farmers (Reddy and Rao 1998; Revathi 1998; AWARE 1998, Choudary 2002; Vidyasagar and Chandra 2004).

Andhra Pradesh has been a leading state in the incidence of farmers' suicides. The first phase of suicides was committed by cotton farmers in the mid nineteen eighties (1986-87) in Prakasam district. The second phase started in the mid nineties (1997-98) beginning with Warangal district, but spread to some other districts of north Telangana with suicides by cotton farmers.

These farmer suicides in Warangal district still continue because of the prolonged drought prevailing in the district. Though there has been some good rainfall in recent years, the lack of proper irrigational projects has led to the inability to store rain water for the cultivation of crops. Though there has been a change in government and even after they have come up with new farmer friendly policies, there is still a prevalence of farmer suicides in the district. During the period of 1999 to 2007, a total number of 677 farmer suicides have been identified in the district so far, according to the District Revenue Office. This dismal condition in the district drew my attention to a study of the agrarian crisis in the district.

1.6. Objectives

The objectives of the study are-

- To study the reasons behind the continuing farmer suicides in Warangal district.
- To study the measures taken by the Government to overcome the reasons that contribute to the continuity of farmer suicides in Warangal district.
- To study the adequacy of the measures.

1.7 Methodology

The present study was conducted in the agrarian state of Andhra Pradesh. It will focus on the dismal state of agriculture in the Warangal district of Telangana region of Andhra Pradesh. The region was selected for study as 2/3rd of the farmer suicides have been noticed in the Telangana region and among which Warangal district has seen the highest rate of famer suicides in the state. The district's agriculture sector has seen some bad times with drought, crop failures resulting from pests and attack from diseases and the non-availability of credits playing their part. By analyzing various aspects of the district data in the agricultural sector like yield, productivity, land utilization, irrigational facilities, agricultural credits etc. the study has been carried out to determine the reasons for the dismal state of agriculture in the district. The primary sources of the data have been gathered from the families of deceased farmers from three mandals of the district namely Ghanpur (stn.), Dharmasagar and Atmakur and the major observation focuses on the farmer suicides which took place during the period of 2000-01 to 2006-07.

1.8 Sources of Data

For the purpose of this study both primary and secondary data were used. The general details of information regarding the agricultural scene in Andhra Pradesh has been sourced from secondary sources like Economic Reviews, Statistics for Planning published by Directorate of Economics and Statistics, Andhra Pradesh, Agricultural Statistics published by the Directorate of Economics and Statistics, A.P, Human Development Index Report, Andhra Pradesh, Statistics for Planning published by the Chief Planning Officer, Warangal, Agricultural Action Plan Reports of the Joint Director of Agriculture Office, Warangal and other relevant magazines and newsletters. Three Mandals of the district, namely Ghanpur (Stn.), Dharmasagar and Atmakur have been selected for gathering primary data from the families of deceased farmers. This primary data has been selected from the list of farmer suicides prepared by the District Revenue Department on the basis of the report filed by the three member committees of the Mandals, appointed by the District Collector. A questionnaire has been prepared for the purposes of study in such a way so as to gather information regarding the families of the deceased based on their socio-economic conditions. Apart from this, important personalities related to the issue, like authorities of both formal and informal institutions and leaders of the farming community have also been consulted to obtain relevant information for the study.

1.9 Organization

This thesis has been organized into five chapters. In the second chapter, an attempt has been made to provide a proper review of the available literature on agricultural problems in the country with an emphasis on studies related to states like Andhra Pradesh, Tamil Nadu, Kerala Karnataka and Maharashtra and literature related to the agricultural situation in Warangal district. The third chapter focuses on the particulars of the area that is the focus of the study. It gives an overview of Andhra Pradesh agriculture and the agricultural features of Warangal district. In the fourth chapter, I have made an attempt to compare the agricultural situation in the district of Warangal with the whole of the Andhra Pradesh economy, through an analysis of the field study information. The fifth chapter is a summary of the entire study, where I discuss the probable remedies and implications of policies.

1.10 Limitations

This thesis does not make any claims of being complete or comprehensive. There are certain limitations to this study related to the use of secondary sources of data which can creep into the framework of the study. But efforts have been made to collect data that is most recent in order to make the study more up-to-date and relevant.

CHAPTER TWO

AGRARIAN DISTRESS: A LITERATURE REVIEW

Agriculture forms a large part of India's economic base. It is treated as the primary sector of the Indian economy. In addition to being an occupation and a business proposition for several people in India, agriculture is also a tradition, a way of life, which for centuries has shaped people's thoughts, outlook and culture.

"Over the past few decades, the Indian agriculture has had remarkable technological changes and there is promise of arrival of even more sophisticated technologies in the years to come. The sector is also believed to have a wide range of commodities enjoying comparative advantage in international trade. Despite these favorable features, the sector is described by many as having become 'a parking lot of the poor' over the planning decades...[1]"

Up until Independence, the Indian government thoroughly neglected the agricultural sector and made no attempts to uplift the farmers from their deteriorating plight of impoverishment and debt. Before Independence, agriculture was practiced only as subsistence farming and it was more a way of life than a profession. Agricultural activities were segregated according to regional differences that arose out of varied physical resource endowments. These differences were also further accentuated by the varying levels of investment in rural infrastructure and in technological innovations that might improve conditions in agricultural areas. This controlled and regulated environment offered very little incentive for resource exploitation.

In June 1991, the Indian Government implemented a new set of economic reforms that were aimed at putting the economy on a high growth path, through improved efficiency in the industrial sector, by bringing in global competition. The continued presence of acute agricultural issues in many Indian states may be primarily due to the loopholes in formulated agricultural policies. A lack of clear understanding about the farmers' requirements may be an important factor that led to the failure of various policy initiatives. Times have changed though; planning has helped the Indian economy reach a stage of development, where further progress can be made through a basic change in the strategy and role of the individual states, in the national economy. The fundamental focus here is how best a state can help prompt growth and ensure that the benefits reach all sections of society.

The Government of India assembled a committee, under the chairmanship of M. Narsimhan in 1991, to review the current financial system of that time. The committee was successful in coming up with several recommendations, which were – a) Elimination of concessional lending rates even for the priority sectors and for government sponsored programmes like IRDP, b) On the structure of rural credit institutions, it suggested the setting up of rural banking subsidiaries, by public sector commercial banks. Apart from these two, there were also other recommendations made by the committee that effected numerous changes in the financial system, particularly in agricultural development.

"Inadequacy of public investment in agriculture is today a matter of general concern. This is an area, which is the responsibility of states, but many states have neglected investment in infrastructure for agriculture. There are many rural infrastructure projects, which have been started but are lying incomplete for want of resources."

Anil Sharma and Ashok Gulati (2004) have inferred that subsidies can serve only as short term measures and cannot be substitutes for long term measures in social and physical infrastructure. Long term measures ensure the viability of new technologies. Gulati and Sharma argue that subsidies trim down the resources available for productive investments that can be utilized in generating new productive potential, especially in a resource crunched country like India, where any particular achievement comes at the expense of other competing objectives. In their opinion, Indian agriculture isn't net subsidized, but net taxed. Subsidies have an adverse effect on fiscal position, cropping pattern, the environment, equity, new industries and technology. Their studies on subsidy arrives at a similar conclusion that for an accelerated and sustainable growth of agriculture, a reform policy package must include liberalisation and removal of agricultural input subsidies, accompanied by targeted safety net programmes and an increased investment in irrigation.

V.M. Rao based his studies on analysing whether agricultural growth could help the poor break through the constraints and barriers that keep them below poverty line. To do this, Rao looked into the characteristics of agricultural growth in India during the planning era. Rao relied on the work of V.M Dandekar (1994), for assessing the type and level of growth.

He was able to infer that the actual performance of Indian agriculture always fell short of the modest targets throughout the decades, and an even more worrisome parameter was the increase in population growth rate. In the liberalised regime, the ability of agriculture to advance was largely dependent on the pace at which the rural population responded and adjusted to the growing integration of traditional village economy with the opened National economy. Over the planning decades, wheat and rice was observed to have benefited most farmers. There had been a substantial increase in irrigation and more over than it had been grown in highly assured rainfall areas. Irrigation methods and technologies improved manifold, especially in low-rainfall regions. The good growth recorded in few areas and of few crops may be the reason for the slight agricultural development, even though there is near stagnation over larger areas. Another decisive parameter was the disparity between agriculture-based and non-agriculture-based income that was reflected in the per capita GDP. Dandekar believed that the disparity arose from the flawed structural features of the Indian economy, and that despite this disparity, the population does not move from agriculture to non agriculture (there still aren't high levels of urbanization).

D.N Reddy and Srijit Mishra (2008) stated that the agricultural distress, as expected, had a greater impact on the small holders with limited resources and the farming community inhabiting the rain-fed regions and tribal areas. They have tried to evaluate whether small marginal farming is sustainable without public infrastructure support and comprehensive social security covering health, education, employment and old age support. According to their study, the public funding for research and education is low in dry land states with harsh environments. The increasing capital intensity of new technology and institutional bias in the delivery of technology restricts access to technology for small and marginal farmers (Suresh Pal). In the post-reform period, high gross domestic product (GDP) growth accompanied by low agricultural growth has brought about a rapid shift in the sectoral distribution of GDP which is skewed against agriculture. The declining share of agriculture is a tendency driven by forces inherent in the development process. However, the paradox is that there is no commensurate decline in the share of agriculture in the country's labour force. Reddy and Mishra have observed low growth of farm productivity, low agricultural prices, slowdown of demand for agricultural products due to the stagnation of per capita food consumption during 1993-2005, and inadequate employment opportunities outside agriculture are the proximate causes that underlie the current slow pace of agricultural transformation.

Non-agricultural liberalisation has been noted to be more beneficial by Narayan, Parikh and Srinivasan (1990), for agriculture, even in spite of agricultural liberalisation will also accelerate growth. In the short run, agricultural trade liberalisation itself disrupts agricultural production due to the prevailing capacity constraint, resulting in a spike in import. But over time, capacity will be increased and imports will stabilize. On the other hand, if protective tariffs are being imposed, it should be done after careful evaluation of benefits and costs. However, there are other studies that strongly recommend high tariff measures for the protection of the agricultural sector. As the comparative advantage in free trade is due to technological differences (Ricardian Theory of International Trade) and relative factor price differences (Hecksher-Ohlin model), the large scale subsidies provided by the government in developed countries makes their agricultural production highly profitable. This study emphasizes the fact that the governments of developing economies have to impose high tariffs on imports of agricultural products, not only to curtail unfair trade practices, but also to transfer a portion of the subsidies into the tariff revenue, which can then be utilized to increase investment in the agricultural sector and improve productivity.

V. Puhazendi and B Jayaraman (1999) say that political intervention, deficiencies in loaning procedures of credit institutions, especially those pertaining to appraisal, supervision and control have contributed to the malady of high overdues.

By studying the working of RRB's in Karnataka, Jayaraman and Birdar's study attempts to understand the problem of overdues and non-repayment of loans disbursed under different schemes from the RRB's. Jayaraman and Birdar(2002) found that about 65% of the beneficiaries had deliberately postponed the repayment of their respective loans with the expectation that the government would waive off their loans in the future. The study also observed that the borrowers did not receive their credit in a timely manner, and that there was a misuse of credit. Owing to complicated procedures followed by the RRB's in sanctioning loans, borrowers were compelled to approach private money lenders instead of availing the banks' schemes.

Dutt and Sundaran (2004) found that one of the major problems faced by lending institutions, particularly the co-operatives, was the high number of overdues. Improper management and regional disparities were also contributory problems faced by the institutions. The health of agricultural credit institutions, both in the Co-operative and Regional Rural Banks, was in a sad state of affairs in several parts of the country. The Planning

Commission issued a statement which said, "Willful default and over dues are mounting in a number of states like Maharashtra and Gujarat.

D.K. Desai (1988) opines that in India, institutional credit for agriculture was provided with the intention of protecting farmers from the guile of moneylenders. His study revealed that the share of institutional credit to the outstanding debt of cultivator households increased from 12.3% in 1951-52 to 63.3% in 1981.

In his study on the alarming increase in farmer suicides in Andhra Pradesh, Chowdary (2004) analysed the reasons behind the suicides. According to him, the major factors leading to agricultural difficulties are, negligence of the agricultural sector by the state and central government, non-remunerative nature of farming, indebtedness of farmers, low quality seeds, overuse of fertilizers and pesticides, declining productivity, lack of efficient irrigation facilities, power supply shortage, inefficient agricultural extension services, diminishing public interest in the agricultural sector, absence of proper agricultural insurance, falling agricultural prices, problems related to tenant farming and extensive export orientation in agriculture.

Mohanty (2004) says that prevailing agro-climatic and socio-cultural features of Maharashtra are different from that of Andhra Pradesh and Karnataka. In Maharashtra, the focus is more on commercial cultivation, with the dominant produce being cotton. The available institutional facilities are however, not up to the mark, making farmers largely dependent on informal sources of credit. At times, decrease in production and sales occur, since new methods of farming render traditional knowledge and skills obsolete. There is a lack of adequate research and extension services are lacking. Marketing methods are also ineffective. Moreover, the dependence of agriculture on the vagaries of the monsoons and other climatic conditions has a detrimental effect on the crops. The study further says that it cannot be conclusively inferred that the existing agricultural situations are the sole reasons for farmer suicides, but they have contributed immensely to the continued occurrence of suicides.

V.S Vyas (1994) points out four major weaknesses that plague Indian agriculture – a) preponderance of low value agriculture, b) low cost-benefit ratio, c) inefficient use of natural resources, and d) deterioration of self-help institutions. These four are considered to be major factors that affect sustainable agricultural growth. Vyas also says that the income levels of Indian farmers are low, because the returns on their produced crops are low due to lack of proper pricing in the markets. The low cost-benefit ratio is mainly due to the inefficient use of resources such as fertilizers, irrigational facilities, etc. Hence, these factors make Indian agriculture, high-cost agriculture. Depletion of natural resources has also been given priority in his study. Though measures have been taken to stop deforestation, cited as an important cause for agricultural land degradation like soil erosion, fall in ground water levels etc., there are other factors that he identifies as contributing to the problems that plague Indian agriculture. One of them is the failure of credit in reaching the agricultural sector, due to delays introduced by the bureaucracy. This flaw causes the farmers to rely on informal financial institutions. Therefore, he suggests that an intensive institutional support with a proper policy backup, in an extensive way, can help overcome the various challenges.

The Green Revolution tremendously increased food production and accelerated the pace of modernization and urbanization in India. On the negative side, it also created serious regional imbalances and increased the gap between classes of farmers. Sutapa Lahiri (2000) argues that when Punjab attained growth in agricultural production, there was no corresponding industrial development. Agricultural development and the Green Revolution therefore reached their saturation point in Punjab. The share of agriculture in the Net State Domestic Product (NSDP) declined to 25.42% in 1997-98 from 34.69% in 1991. In 1951, 55.89% of the labour force in agriculture produced 48.62% of Punjab's NSDP, whereas the same proportion of labour in 1991 produced only 34.69% of the NSDP. Share of cultivations to total workforce declined, displaying a significant increase in the proportion of landless agricultural labourers. Soil quality deteriorated, which led to the requirement of more fertilizers. Punjab's agriculture then became a mono-crop agriculture one, rather than being diverse. The growing pressures of population on land, low productivity, diminishing economic returns, poorly organized marketing facilities, declining public investments, technological constraints, and indebtedness made agriculture no longer an attractive option in Punjab. The reduction in both export subsidies and direct subsidies as per the WTO agreements, given the downward trend of production and stagnating yield of agriculture in Punjab further hampered its agricultural competitiveness. Lahiri's study suggested solutions like diversification of agriculture in high value crops and the development of a vibrant and dynamic rural non-farming sector.

The study conducted by Rao and Gopalappa (2004) mainly focuses on the inadequacies of the present policy regime in Karnataka. Despite the fact that large parts of the state are drought prone areas, there has been a substantial shift towards the commercialization of agriculture. The policy regime has been such a non-caring one and it has not given any importance to the goals of enabling growth and modernization of agriculture with adequate benefits to farmers. The policy regime should be considerate, in terms of strengthening its foundation, growth prompting services and safety nets.

Bhowani Sen (2002) states that an inherent difference between industry and agriculture is that the latter is dependent on weather conditions. The application of science in agriculture through the use of astronomy and weather control has many limitations. Under the first two Five Year Plans, the rate of improvement in agriculture produce was only 3.3% per annum at a cost of Rs.1500 crore and the country was compelled to import food grains at the approximate rate of 3 million tonnes per year. Productive resources pumped into the rural economy tend to be nabbed by vested interests and so the vast mass of tillers of the soil, who alone are able to increase production, are largely deprived of these benefits. Sen's study focuses on Maharashtra and West Bengal, the two most industrialized areas, and puts forward the fact that, in spite of various land reform acts, there exists in these states a preponderance of inferior tenants and crop sharers, being 30% and 22% respectively. Bhowani suggests that cooperative movements, with proper government support and aids can act as the peasants' primary weapon against the monopolies in rural areas, and will definitely help in the regeneration of the rural economy.

The study by R.V Dadibhavi (1998) has attempted to examine the regional disparities affecting the flow of institutional credit to agriculture. Dadibhavi's study was based on the following objectives – a) to analyse the regional disparities in the disbursement of institutional credit to farmers, b) to examine whether commercial banks venture into areas where co-operatives are also strong, and c) to analyse the distribution of credit according to the size of holdings of the borrowers. The study examined regional disparities in institutional credit, in terms of loan outstanding per hectare of net sown area, provided by all credit institutions. The study revealed that, on the basis of total loan outstanding per hectare (Co-operatives + commercial banks) in 1982, Kerala was on the top of the list with Rs. 2,486 crore outstanding credit, followed by Tamil Nadu with Rs.1,369 crore, Punjab with Rs.740 crore and Himachal Pradesh with Rs.720 crore. The study has suggested that the identification of credit deficit states may serve as a guide for the branch expansion policy of commercial banks and insists that they can still enter the areas where co-operatives are known to be financially strong.

Indian agriculture is in a state of flux. Market driven diversification in a global perspective has become the new paradigm that is driving future agricultural growth. Some high-value sectors such as livestock, horticulture and fisheries have been responding to these market forces, while the crops sector continued to stagnate in the last decade or so. There is a consensus that policies and instruments that have served well in the past are either overstretched or are untenable under this new scenario (Suresh Pal). Studies carried out by S.Galab, E.Revathi and P. Prudhvikar Reddy have revealed the manifestations behind the agriculture crisis in Andhra Pradesh, of which the aspect of farmer suicides is one that comes as a rude jolt. These authors have thoroughly discussed the countrywide attention that Andhra Pradesh garnered on account of being the state that recorded the highest spate of farmer suicides. According to their study, symptoms of the deep malaise of the agrarian crisis, is a result of a combination of factors that include the growing process of marginalization in agrarian structure, the increasing fragility of land and water resources and sustainable cropping practices, all of which were aggravated by the neglect of public support systems due to the economic reform process.

A citizen's report titled 'Gathering Agrarian Crisis-Farmers' suicides in Warangal District of Andhra Pradesh in 1998' by A. Sudarshan Reddy, observed that a common feature that has emerged on the agrarian scene over time, both at the state and district level, is the proliferation of small farms. Since the formation of Andhra Pradesh to date, the number and percentage of small farms that measure less than 5 acres has gone up by about 80 percent. Holdings of less than 2.5 acres constitute more than 50 percent and are growing at a rapid rate. Their share of land has also increased considerably. The factors contributing to such a trend include the growth of the population, subdivision and fragmentation, lack of opportunities outside agriculture, decline in caste occupations, break down of the joint family system and in a small measure, land reforms etc. From the point of view of farm resources, it will enhance labour supply and thus provide scope for increased agricultural production through intensive cultivation. This can be effective provided that other resources are being ensured on the farm. However, there are a few disquieting features that can be seen in the stagnating agricultural work force on the one hand and the declining share of cultivators as compared to agricultural workers on the other. The plausible reasons for this trend appear to be the declining share of agriculture in the state domestic product without a corresponding reduction in the agricultural population. The other reason is possibly the inability of very small farms to generate adequate income and employment to sustain themselves as cultivators.

CHAPTER THREE

ANDHRA PRADESH-AGRICULTURE: A DESCRIPTION

3.1 Introduction

Andhra Pradesh was one of the first states in India to be formed only on the basis of language. After India's independence, the Andhras, that is, the Telugu-speaking people (although Urdu is widely spoken in Hyderabad) were distributed in about 21 districts. Out of them 9 were in the Nizam's Dominions and 12 in the Madras Presidency. But after an agitation, on October 1, 1953, 11 districts of the Madras State were consolidated to form a new Andhra State with Kurnool as the capital. Later, on November 1, 1956 the State Reorganization Commission recommended the enlargement of the Andhra State and added nine districts that were in the Nizam's Dominion. The city of Hyderabad, which was also at the time the capital of the Nizam's dominion, became the capital of the enlarged Andhra Pradesh.

On 1 November 1956, Andhra State merged with the Telangana region of Hyderabad State to form Andhra Pradesh. Hyderabad, the former capital of the Hyderabad State, was made the capital of the new state, Andhra Pradesh. The 23 districts of Andhra Pradesh can be divided into three regions, namely, coastal Andhra, Rayalaseema and Telanagana.

3.2. Population Growth Rate

Population trends of Andhra Pradesh:

Census	population	%
1961	35,983.000	-
1971	43,503,000	20.9%
1981	53,550,000	23.1%
1991	66,508,000	24.2%
2001	75,727,000	13.9%

Source: Census of India.

3.3 Economy

Agriculture has been the chief source of income for the state's economy. Three important rivers of India, the Godavari, Krishna and Tungabhadra flow through the state and provide an important source of irrigation. Rice, Sugarcane, Cotton, Mirchi (chilli pepper), Mango and Tobacco are the local crops. Recently, crops used for vegetable oil production such as sunflower and peanuts have also gained favour. There are many multi-state irrigation projects in development, including the Godavari River Basin Irrigation Projects and the Nagarjuna Sagar Dam, the world's highest masonry dam.

Andhra Pradesh is a mineral rich state, ranking second in India in terms of mineral wealth. The state has about one third of India's limestone reserves, estimated at about 30 billion tonnes. The Krishna-Godavari basin contains huge reserves of natural gas and petroleum. The state is also blessed with huge amount of coal reserves.]The state ranks first nationwide in hydro electricity generation with a national market share of over 11%.

Andhra Pradesh's GSDP for 2005 was estimated at $62 billion in current prices. This is a chart of the trend of Andhra Pradesh's GSDP at market prices estimated by the *Ministry of Statistics and Programme Implementation* with figures in millions (Indian Rupees). Accordingly, the state ranks fourth in terms of overall GSD and fourth in per capita GSDP among the major states of India.

Year	State GDP (Rs. MM)
1980	81,910
1985	152,660
1990	333,360
1995	798,540
2000	1,401,190

Source: Census of India

3.4. Land Utilisation

The word "Land" is used in Agricultural relativity, with reference to the cropped or cultivated land. It is a precious resource provided by nature to the state.

Land Utilisation in Andhra Pradesh:

(Lakh. ha.)

Item	97-98	98-99	99-2000	2000-01	2001-02	Average of 1997-2001
Total Geographical area	274.40	274.40	274.40	274.40	274.40	274.40
Gross cropped area	121.35	136.25	130.27	135.45	127.56	130.18
Net cropped area	89.46	109.78	106.10	111.15	104.10	104.12
Gross Irrigated area	51.58	60.92	57.46	59.16	55.49	56.80
Net irrigated area	39.44	45.39	43.84	45.28	42.38	42.73

Source: Director of Economics & Statistics, Hyderabad.

3.5. Soils

Andhra Pradesh is endowed with a variety of soils ranging from poor coastal sands to highly fertile deltaic alluviums. Red soils occupy over 66% of the cultivated area and are mostly situated in the Rayalaseema districts. These soils have a low nutrient status. Red earth which are commonly termed as Red soils can be sub-classified as (a) Dubba soils (loamy sands to sandy loams) (b) Chalkas (Sandy loam soils) (c) Sandy clay loams (d) Loams including silty soils (e) Deep loamy sands and (f) Sandy loams with clay subsoil. Chalkas occur mostly in the Telangana districts while red loams combined with sands are present in the upland regions of coastal districts. Black soils cover nearly 25% of the cultivated area and are generally associated with poor drainage. They are also called Regurs or Vertisols and are of two types. The first category is in-situ soils while the second is transported soils. While the first category can be found in the Coastal districts and parts of Telangana and Rayalaseema, the second category occurs in the valley regions of the slopes with calcareous concentrations. The in-situ soils are generally heavy in texture and high in salt concentration. The alluvial loamy clay soils found in Krishna and Godavari deltas cover 5% of the cultivated area. The coastal sands occupy only 3% while the remaining 1% is covered by laterite soils in certain pockets of the State.

3.6. Climate

The state has a tropical climate with moderate diffusion to sub-tropical weather. Humid to semi-humid conditions prevail in the coastal area while arid to semiarid situations are pronounced in the interior parts of the state, particularly Rayalaseema and some districts of Telangana. The areas covered by the Deccan plateau are characterised by hot summers with relatively pleasant winters.

3.7. Rainfall distribution

The state as a whole benefits from both monsoons but Telangana gets less than 25 % of the total N.E. Monsoon. The annual rainfall varies from less than 60 cms in Western Rayalseema to more than 100 cms in the North. The N.E. part of the state, close to the north eastern border that crosses the Eastern Ghats reaches as high as 150 cms. In the Telangana region, rainfall varies from 75 cms in the Southern half to over 100 cms in the Northern half. More than 75% of the rain is received during the S.W. Monsoon i.e., June to September. An in-depth study of the behaviour of monsoons over a period of years shows that the general characteristics of rainfall in this region is erratic with a continued dry spell of three weeks or more thus creating drought conditions. Heavy spells are generally noticed in July and September resulting in two fillings of tanks and reservoirs.

Region and season wise rainfall particulars of Andhra Pradesh:

Monsoon	Coastal Andhra	Rayalaseema	Telangana	Andhra Pradesh
South West Monsoon				
Normal	620	407	715	624
2005-06	665	418	820	690
2006-07	649	307	734	627
2007-08	804	647	737	747
North-East Monsoon				
Normal	324	238	129	224
2005-06	439	432	172	321
2006-07	288	157	64	168
2007-08	250	227	60	163

Source: Directorate of Economics & Statistics, A.P.

The cropped area in Andhra Pradesh is divided into seven zones based on their agro-climatic conditions. The classification mainly concentrates on the range of rainfall received and the type and topography of soils. The districts covered by the different zones and their agro-climatic characteristics are given below:

A. Krishna-Godavari Zone

B. North Coastal Zone

C. Southern Zone

D. Northern Telangana Zone

E. Southern Telangana Zone

F. Scarce Rain Fall Zone of Rayalaseema

G. High Altitude Tribal Areas

3.9. Cropping Pattern

The state has definite cropping seasons based on the rainfall received through two monsoons that supplement other irrigation sources. They are the Kharif, Rabi and Summer seasons. The cropping pattern mostly depends on the length of availability of irrigation water and the assuredness of such a source of water supply. The following are some of the major cropping patterns adopted round the year as per different water sources.

Major Projects
Medium Projects
Lift irrigation systems
Minor Irrigation Sources (Tanks)
Wells & Tube Wells (assured water supply)
Rain fed – Uplands

Percentage of Irrigated Area by Source across Districts of Andhra Pradesh: (2004-05) (% in Net Sown Area)

S.No	District	Canal	Tanks&Others	Wells	Total
1	Srikakulam	33.2	22.1	4.0	59.3
2	Vizianagaram	12.0	26.6	5.5	44.1
3	Vishakapatnam	11.2	16.4	4.6	32.1
4	East Godavari	42.2	8.2	12.3	62.8
5	West Godavari	40.4	5.8	30.3	76.5
6	Krishna	42.4	7.4	11.3	61.1
7	Guntur	42.0	2.8	8.8	53.6
8	Prakasam	7.2	5.0	13.6	25.9
9	Nellore	20.2	21.4	27.5	69.0
10	Kurnool	7.3	2.1	10.3	19.7
11	Anantapur	2.2	0.2	8.1	10.5
12	Cuddapah	5.1	0.6	23.8	29.6
13	Chittoor	0.0	4.2	27.2	31.3
14	Rangareddy	0.3	1.1	20.7	22.1
15	Medak	0.4	1.8	23.0	25.2
16	Mahabubnagar	2.5	0.8	15.3	18.7
17	Nalgonda	10.2	2.6	23.3	36.2
18	Warangal	1.4	9.7	52.7	63.7
19	Khammam	6.6	16.2	16.8	39.6
20	Karimangar	0.1	2.3	48.5	50.9
21	Nizamabad	1.5	3.8	50.3	55.6
22	Adilabad	0.9	1.8	10.3	13.1
	Andhra Pradesh	**12.9**	**6.1**	**18.2**	**37.2**

Source: Directorate of Economics & Statistics, A.P.

3.10. Agricultural Crisis in Andhra Pradesh

The extreme agrarian distress in Andhra Pradesh achieved national prominence when it resulted in the dramatic increase in suicides by farmers in the region. These have occurred in substantial numbers for about seven years now. The awareness of the severity of problems in the countryside has made agricultural regeneration the most important priority. It is becoming increasingly evident that the difficulties confronting agriculture in Andhra Pradesh are complex and multifarious, and will in fact require a complete reversal of the earlier economic strategy followed in the state. (Jayati Ghosh and C.P Chandrasekhar).

But past experience in Andhra Pradesh deserves even greater national attention. This is because what has already happened in very acute form in this state is occurring in many (even most) other parts of rural India. The same symptoms - farmers' suicides, hunger deaths in the midst of production surpluses, distress migration under stark conditions - are exhibiting themselves in regions as disparate as Vidarbha in Maharashtra and south-western Rajasthan. The causes, also, are broadly the same, with the shift towards increasingly unreliable cash crops, the decline in institutional credit, the problems with input supplies and crop marketing, and the lack of alternative non-agricultural income opportunities, all contributing to the generalised agrarian crisis. Of course, what happened in rural Andhra Pradesh has been a particularly severe case. The state of Andhra Pradesh had become almost a laboratory for every extreme form of neoliberal economic experiment, with a massive shift towards relying on incentives for private agents as opposed to state intervention and regulation of private activity, in virtually all areas. The problem of farmers' suicides, for instance, is probably the most dramatic sign of extreme despair and hopelessness ranking close to starvation deaths as the most blatant indicator of the extent of rural devastation. The proximate cause of such suicides is usually the inability to cope with the burden of debt, which farmers are unable to repay themselves. In most (but not all) cases, the debt is contracted to private moneylenders, as the massive decline in agricultural credit from banks and co-operatives has reduced the access, especially of small cultivators, to institutional credit.

Public agricultural extension services have all but disappeared, thus leaving farmers to the mercy of private dealers of seed and other inputs such as fertiliser and pesticides. These private dealers function without adequate regulation, creating problems like wrong crop choices, excessively high input prices, spurious inputs and extortion. Public crop marketing services have also declined in spread and scope, and marketing margins imposed by private traders have therefore increased. All this happened over a period when farmers were actively encouraged to shift to cash crops, away from subsistence crops which involved less monetised inputs.

The crisis in water and irrigation sources can also be traced to these cultivation patterns. Over-use of groundwater - once again resulting from the absence of public regulation or even advice, as well as the shift to more water-using crops - has caused water tables to fall across the state. Declining public investment, inadequate maintenance and the regionally uneven pattern of spending, have all made surface water access equally problematic. As a consequence, there are now real problems with respect to even the current economic viability of farming as a productive activity in most parts of rural Andhra Pradesh, not to mention its sustainability over time. Other factors have added to debt burdens that become unbearable over time. Production loans dominate in current rural indebtedness. But among the non-productive loans incurred by rural households, those taken for covering medical expenses are the most significant. The deterioration of public health services and the promotion of private medical care have dramatically increased the financial costs of mere physical survival and well-being.

This entire process is sometimes presented as a situation in which rural people have been "left out" of the process of globalisation, or have been "marginalised" or "excluded". But nothing could be further from the truth. The problem is not that cultivators and workers in this state have been "left out"; rather, they have been forced into market relations that intrinsically work against them. They have not been marginalised and excluded; instead, they have been incorporated and integrated into market systems in which their lack of assets, poor protection through regulation and low bargaining power operate to make their material conditions more adverse.

Thus, there have been policy errors of both omission and commission. And these have combined to create the specific problems outlined above, about which there now exist a wealth of data and a number of useful studies. But the effects of these errors are apparent even in the aggregate data on output, even though these are recognised to be notoriously slack in identifying actual material trends. Chart 1 indicates the behaviour of the index numbers for per capita income (that is the net domestic product in constant 1993-94 prices) for all sectors and for agriculture alone. The SDP has been deflated by total population, and the agriculture product has been deflated by rural population only, to give an idea of the actual trends over time in rural incomes.

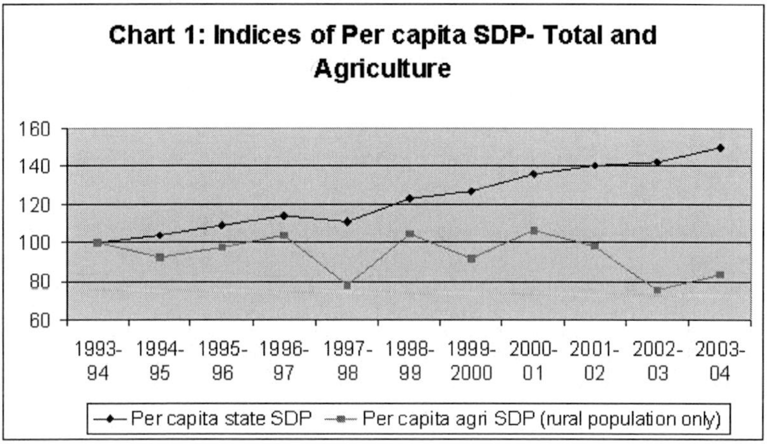

It is evident that while aggregate SDP per capita has increased moderately since 1993, SDP in agriculture per capita shows no such increase, and has actually declined. In fact, between the triennium 1994-94 to 1995-96 and the triennium 2001-02 and 2003-04, per capita agricultural product actually declined by around 12 %.

This has also been reflected in indicators of per capita consumption, which probably provide a more accurate picture of the real economic conditions in the countryside. Chart 2 indicates the trend in the four regions of rural Andhra Pradesh according to the NSS consumer expenditure surveys.

Aggregate per capita consumption for the rural areas of the whole state taken together increased marginally between 1983 and 1999-2000. But it is notable that there appears to have been hardly any increase since 1993-94, despite the moderate increase in per capita SDP indicated above. What is even more significant is that per capita consumption fell after 1993-94 in all the regions of rural Andhra Pradesh barring the coastal Andhra region. This fall was particularly marked for Rayalseema (comprising the Southwest and Inland Southern regions). So, in most of the rural areas of the state, average consumption expenditure actually declined in real terms in the period 1993-94 to 1999-2000.

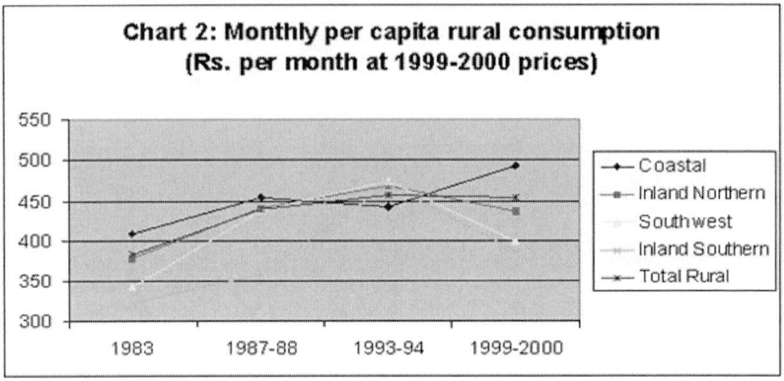

This is quite consistent with the picture of the growing difficulty of cultivation that has been outlined above. But in addition to the agricultural patterns, the general stagnation of the rural economy and the absence of non-agricultural income generation possibilities contributed further to the deterioration of living standards in the countryside. Part of the problem in employment generation stemmed from agriculture itself. Not only was this sector depressed for instance, but its increasing mechanisation implied falling labour use per hectare of cultivation.

Growing mechanisation of agricultural activity was a phenomenon noticed across rural India. In Andhra Pradesh however, this was actively encouraged by the state government, which provided subsidies to large farmers in its attempts to bring a change in cultivation practices through the use of more mechanised techniques. To provide an idea of the scale of these subsidies, the subsidy expenditure incurred by the state government for farm mechanisation increased from Rs. 250 crores in 2001-02 to as much as Rs. 933 crores by 2003-04.

Unsurprisingly, in this context, agricultural employment stagnated. Table 1 provides evidence of the growth of employment, using work participation rates (for all workers and for principal and subsidiary activities) from the NSS and population data from the Census of India. Total agricultural employment in terms of Usual Status occupation barely indicates any increase between 1993-94 and 1999-2000, while in terms of Daily Status (which as a flow measure is a more accurate indicator of labour demand conditions) agricultural employment actually declined.

Agrarian crisis in Andhra Pradesh:

Table 1: Annual rates of growth of rural employment

per cent per annum	Total Usual Status	Total Daily status	Agri Usual Status	Agri Daily Status
1983 to 1987-88	1.04	2.01	0.07	1.53
1987-88 to 1993-94	3.15	83.73	3.66	4.39
1993-94 to 1999-2000	0.06	80.03	0.55	-0.03

The decline in agricultural employment was not countered by any increase in non-agricultural employment, as a result of which total rural employment effectively did not increase at all. This occurred in a period when rural population was increasing in Andhra Pradesh by around 1.4 per cent per year. Andhra Pradesh displayed the lowest rate of rural employment growth (of all types - that is principal and subsidiary) among all the states in India, at less than one-tenth of the already low national average of 0.66 per cent per year over this period.

This too was related to national and state government policies - in particular, the decline in public expenditure directed towards rural areas, especially after 1992-93. The positive multiplier effects of public expenditure in rural India have been widely noted. Indeed, it was such expenditure that was responsible for the diversification of employment observed across the rural economy in the late 1980s, and the associated reduction in rural poverty. In contrast, the 1990s were marked by a concentration of public resources towards urban areas.

This trend was especially marked in Andhra Pradesh, where the urge to build a supposedly modern metropolis in Hyderabad and to enhance certain types of connectivity especially for urban areas, dominated over the evident need to provide resources even for basic infrastructure in villages. Thus, even public investment that directly contributes towards agricultural growth, whether in the form of road and rail connectivity or the construction and maintenance of surface irrigation schemes, was reduced. The pattern of public expenditure therefore contributed in no small measure to the overall agrarian crisis.

District-wise Estimates of Suicides of Farmers in Andhra Pradesh: 1998-2006:

District	1998	99	2000	01	02	2003	2004	2005	2006	Total
Srikakulam		1				1	10	2		14
Vizianagaram							9			9
Vishakhapatnam				1			22	11	2	36
East Godavari			1	2		1	21	3	5	33
West Godavari	1			5	1	2	26	1	2	38
Krishna	4	1	1	3	7	3	52	15	8	94
Guntur	8	22	2	7	4	1	103	27	12	186
Prakasam	1	3		2	3	1	62	9	2	83
Nellore							38	9	5	52
Coastal Andhra										545 (12.37%)
Chittoor	3			2	3	3	51	9	15	86
Cuddapah				4	8	12	39	9	7	79
Kurnool	4	4	2	4	3	18	102	25	49	211
Ananthapur			51	97	78	104	176	25	24	555
Rayalseema										931 (21.14%)
Medak	15	3	2	8	2	4	139	44	39	256
Mahbubnagar	6	21	36	26	41	46	203	63	25	467
Warangal	69	100	98	119	123	92	88	51	26	766
Nizamabad	9	1		11	5	4	122	43	20	215
Adilabad	9	8	5	13	4	13	67	52	27	198
Karimnagar	31	10	6	30	20	16	216	96	32	457
Khammam	20	5	3	6	6	7	49	23	16	135
Nalgonda	5	1	10	11	18	12	142	59	33	291
Ranga Reddy	5		3	6	2	7	58	41	20	142
Telangana										2927 (66.47%)
AndhraPradesh	190	180	220	357	328	347	1795	617	369	4403 (100%)

Source: Compiled by Andhra Pradesh Rythu Sangham (APRS), 2007.

In addition, as has already been noted, certain critical public organisations that provided direct assistance to farmers were either closed down (such as the AP Irrigation Development Corporation, which provided assistance in groundwater exploitation) or greatly debilitated (such as the AP Seed Corporation).

All this is important for an analysis of how the direction of policy must change in Andhra Pradesh, and the good news is that the state government already appears to be aware of much of this. But it also serves as a warning for the whole of India. Indeed, just as Andhra Pradesh provides a dismal example of the disastrous effects of neoliberal economic policies upon agriculture, it may also, in the future, point to alternative policies that allow for viable and sustainable agriculture to develop in India.

CHAPTER FOUR

AGRARIAN SITUATION IN WARANGAL DISTRICT

4.1 Introduction

The district of Warangal lies between the Latitude of 17-19' and 18-36' North and Longitudes of 78-49' and 80-43' East, and is above mid sea level by 870 ft. The geographical area of the district is 12,846 sq. kilometers. The district possesses interesting geographical formations and contains minerals of economic importance. The soils of the district are comprised of sandy loams with patches of shallow black cotton soils, and places that are composed of even medium and deep black cotton soil. The district generally tends to be dry and there is not much fluctuation in its temperature. Warangal gets quite warm during the summer months of April, May and June and continues to be warm through the rest of the year except during December and January, when the temperature drops slightly. The maximum temperatures have been recorded as 50.5 degrees centigrade and 13.5 degrees centigrade respectively.

4.2. Rainfall

The rainy season is brought on in the district with the onset of the South-West monsoon in the latter part of June and ends in the month of September with the closure of the South- West monsoon. The normal annual rainfall of the district is 994.001mm. The majority of it occurs in the months of July, August, and September. About 80% of the rainfall is received from the South-West and 20% of rainfall from the North-East monsoons in the district.

During 2007-2008, up to the end of the third week of September, about 96% of rainfall was recorded

Month	Normal mm.	Actual mm.
October(2007)	86.4	63.5
November(2007)	27.2	10.5
December(2007)	6.0	0.0
January(2008)	5.0	0.0
February(2008)	5.9	23.8
March(2008)	13.20	139.0
April(2008)	15.0	23.3
Total	**158.7**	**260.1**

(Source: Joint Director of Agriculture, Warangal)

Rainfall status in Mandals (2007-08)

Status	No. of Mandals
Excess: (-20% above)	24
Normal: (+20% to -20%)	13
Deficit: (-21% to -59%)	9
Scanty: (-60% to -99%)	5

(Source: Joint Director of Agriculture, Warangal)

4.3. Agro-Climatic Zones

The district falls under the central Telangana Agro Climate Zone that covers all the 50 Mandals based on their climate as characterized by rainfall ranging from 810 to 1023 mm and maximum temperature ranging from 28 to 37 ^{0}C.

4.4. Irrigation

The important irrigation sources in the district are the Ramappa, Ghanpur(m), Pakhal, Lakhnavaram and Salivagu projects under which a considerable area is irrigated. The other sources of irrigation in the district are rain fed tanks, wells and hill streams which require good showers in the season. Though the district has perennial and mighty rivers flowing through the surrounding areas, its irrigational sources are still inadequate. The available irrigational sources are not enough to cultivate the available land in the district. Farmers continue to depend mostly on rain water and on bore wells. This may be on account of the geographical location of the district which makes it difficult to create irrigation projects here. Studies say that for the altitude of the district and its ground areas, lift irrigation is the most suitable way of providing irrigational provisions from those flowing rivers. Even now, many projects that were projected to be finished are still under construction and are yet to be finished even after many years.

4.5. Demographic Particulars

With a total population of 32.46 lakhs (2001 census) the district occupies 13th place in the state of Andhra Pradesh in terms of its population and 12th place with regard to its area. There are 262 persons per sq.km in the district as against 277 in the state. As a whole, 22.49% of its population resides in urban areas.

Out of 1098 revenue villages of the district, 1003 villages are inhabited and the rest are deserted. The total rural population of the district is 26.23 lakhs. There are 5 towns in the district viz. Warangal, Kadipikonda, Jangoan, Mahabubabad and Dornakal. The total urban population is 6.2 lakhs. Warangal city is classified as such by the Municipal Corporation.

The working class population in the district is 15.66 lakhs, which is 55.53% of the total population. The workers in the agricultural sector are 9.14 lakhs, forming 28.12% of the total district population. The remaining 4.34 lakhs are workers engaged in non- agricultural sectors comprising 27.7% of the total population of the district. The non working class population is 25.52 lakhs, constituting 78.62% of the total district population.

4.6. General Statistics

As per the 2001 census, the district statistical data is as follows

Sl.no	No. of villages in the district		
1	Revenue divisions	Nos	5
2	Agricultural divisions	Nos	9
	No. of villages	Inhabited	1003
		Un habited	95
		Total	**1098**
3	No. of Gram Panchayats		1013
4	No. of Mandals		51
5	Population	Male	16,37,682
		Female	15,93,492
6	No. of holdings	Marginal farmers	3,14,469
		Small farmers	1,16,368
		Big farmers	75,835
7	No. of agricultural labourers	Male	2,21,879
		Female	3,15,355
		Total	**5,37,234**
8	No. of literates	Male	9,90,384
		Female	6,42,484
		Total	**16,32,868**

(Source: Chief Planning Office- Warangal)

4.7. Land utilization

The word "Land" is used in Agricultural relativity, with reference to the cropped or cultivated land. It is a precious resource provided by nature to the state. The land utilization of Warangal District comprises different natural vegetation levels like forest, barren lands, grazing lands, cultivable lands etc. Most of the land proportion is under cropped lands and forest. Still, the uncultivable land proportion can also be observed in the District land utilization particulars.

Land Utilisation of Warangal District

Area in hct.

	2000-01	2001-02	2002-03	2003-04	2004-05	2005-06	2006-07
forest	371245	371313	371313	373640	373640	373603	371014
barren	53426	53797	53797	54425	54426	54026	52148
non agri	57964	59058	59058	58690	58690	58782	59976
Cultvbl waste	34072	34081	34081	34068	34068	33396	11180
grazing	50400	50401	50401	50917	50917	50917	45260
misclen	3874	3876	3876	3874	3870	3870	50293
cur fallow	107390	107395	187114	577454	148505	110024	115659
other fallows	112834	112834	114809	130484	113237	109983	108094
net area sown	490653	490796	409103	467257	447232	488950	470976
Tot geo area	1283555	1283551	1283551	1283552	1283552	1283551	1284600
Total cropped	508223	593933	508332	508322	109101	612948	609846
area sown once	101317	103137	99229	99229	102413	123989	138870

(Source: Hand book of Statistics, Warangal)

4.8. Area, Yield & Production of Principle Crops of Warangal

As most of the cultivation in the district depends upon monsoons rather than other irrigational sources, variations in the shift of crops can be noticed in the district cultivation every year. The principle crops of the district are Paddy, Jowar, Groundnut, Cotton etc. If we study the area, yield and production details of the principle crops in the district, it can be inferred that Paddy and Cotton play a vital role in the cultivation levels.

Area, Yield and Production of Principle Crops of Warangal District

(in 000 hect./000 Quintals)

	Paddy			Jowar		
	Area	Production	Yield	Area	Prod.	Yield
Year						
2002-03	135	293	2165	11.94	12	652
2003-04	170.23	511	3000	12	14	762
2004-05	130	356	2730	10	11	597
2005-06	202	646	3201	9	5	524
2006-07	209	589	2821	9	5.2	530
2007-08	187	559	2996	7	5	598
	Groundnut			Cotton		
	Area	Production	Yield	Area	Prod.	Yield
Year						
2002-03	40	38	948	119	169	241
2003-04	45.06	58	1293	123.55	279	384
2004-05	45	52	1152	153	288	320
2005-06	42	61	1459	151	372	418
2006-07	37	52	1378	154	359	397
2007-08	37	56	1496	160	475	504

(Source: Directorate of Economics and Statistics, A.P)

4.9. Soil Fertility Status

In Warangal District, Red soils constitute 44% of the total with an area of 5, 65,148 hec. The Black soils constitute 15% with an area of 1, 91,886 hec. Loamy soils constitute 11% with an area of 1, 43,856 hec. and Sandy loam soils constitute 7% with an area of 92,790 hec.

The soil fertility status of the district is as follows

Sl.no	Fertility Status	No. of Mandals in the Status		
		Low	Medium	High
1	Organic Carbon(N)	0	36	14
2	Phosphorus(P_2O_5)	4	38	8
3	Potash(K_2O)	0	0	50

(source: Joint Director of Agriculture, Warangal)

From the analytical reports on the district, it can be observed that Zinc is more deficient than other micronutrients like Ferrous, Magnesium and Boron. Thus, farmers in the district are being advised to apply $ZnSO_4$, along with chemical fertilizers as per the recommendations furnished in the soil test analytical data which also includes recommendations about bio fertilizers and green manures.

4.10. Farming Situations in the District

The problems connected with production and productivity of crops in the district tend to vary according to the micro climates of that particular region. As such, farming situations mentioned below have been identified in the district as suggested by the DAATTC, Warangal, in order to improve the production and productivity.

Farming Situations in the District

Sl.No.	Farming Situations	No. of Mandals
1	Rainfed- Red soils	All Mandals
2	Rainfed-Black soils	-do-
3	Well Irrigated-Red soils	-do-
4	Well Irrigated-Black soils	-do-
5	Tank Irrigated-Red soils	-do-
6	Tank Irrigated-Black soils	-do-
7	Tank Irrigated-Problematic soils	8 (Chityal,Narsampet,Chennaraopet, Bhoopalpally,Govindaraopet, Mangapet, Ghanpur(Mlg) & Venkatapur)
8	Canal Irrigated-Red soils	6 (Hasanparthy, Parkal, Regonda, Shayampet, Duggondi & Mangapet)
9	Canal Irrigated-Black soils	5 (Hasanparthy, Parkal, Shayampet, Duggondi & Mangapet.)

(Source: Joint Director of Agriculture, Warangal)

4.11. Agriculture Situation in Warangal

Agriculture in Warangal district includes the cultivation of food crops, Horticulture, Sericulture, Animal Husbandry, Fisheries and other allied activities and is the source of livelihood for above 72% of the district's population. The socio-economic condition of 87% of the farmers is weak. The effects of drought and floods further weaken their livelihood. The minimum farm produce accrued along with the circumstances detailed above, many a times, fails to bring economic returns due to transport and marketing constraints. Other social drawbacks like poverty, ill health and commitments like marriages and so on add to their problems.

4.12. Agriculture crisis

Agriculture in the district has witnessed adverse conditions due to various natural calamities and institutional failures. In the recent decade the agriculture sector, the major source of occupation and income of the district population, has been worst affected by continuous drought and untimely rains. The scarcity of proper irrigational facilities and institutional support like provision of credit along with continuous crop failures and indebtedness has resulted in agricultural distress in the district, thereby causing farmer suicides in shocking numbers. Warangal district has been identified in the recent decade as one of the districts, not only in the state but in the country that has recorded the highest farmer suicides. According to the official reports of the district's Revenue department, the details of farmer suicides committed during the period between the years 1998 to 2007 are as follows. The following table shows the year wise and month wise statistics of farmer suicides in the district in the recent decade.

Year/Month wise farmer suicides in which exgratia paid by collector as on 25-05-2009

	1998	1999	2000	2001	2002	2003	2004	2005	2006	2007	**Total**
Jan	0	4	9	7	9	4	3	8	1	1	**46**
Feb	0	5	9	4	7	12	8	8	1	1	**55**
Mar	1	14	1	9	12	12	6	6	0	1	**62**
Apr	0	9	3	4	11	6	8	1	2	2	**46**
May	0	6	4	6	10	7	15	1	1	1	**51**
Jun	0	2	3	10	2	3	6	3	0	1	**30**
Jul	10	5	1	7	6	4	17	2	2	0	**54**
Aug	5	4	3	7	10	4	21	1	1	0	**56**
Sep	7	8	15	12	6	12	16	4	4	0	**84**
Oct	6	9	8	8	10	2	11	1	1	1	**57**
Nov	10	7	11	9	3	7	9	4	4	2	**66**
Dec	6	5	11	14	11	3	15	3	0	2	**70**
Total	**45**	**78**	**78**	**97**	**97**	**76**	**135**	**42**	**17**	**12**	**677**

(Source: District Revenue Office, Warangal)

4.13 Suicide relief payments

The "Three Member Committee" constituted by the District collector, after carrying out a vigorous survey on farmer deaths in all the five divisions of the district have identified that 677 farmer suicides were due to agricultural distress. When suicides have been officially certified by the committee, the affected families are paid an exgratia so as to facilitate them with financial support to overcome their distress to some extent. The table given below shows the details of exgratia financial support, released to families affected by farmer suicides that have been identified by the division committees. So far, only a certain portion of the amount has been released to a few of the affected families. There are still families that are yet to be facilitated by this financial support.

Farmer's suicide relief payment details as on 25.05.2009 (amount in Lakhs)

Sl.No.	Division	No. of cases	Amount
1	Warangal	252	378.00
2	Mulug	192	288.00
3	Mahabubabad	76	114.00
4	Narsampet	50	75.00
5	Jangaon	107	160.50
	Total	**677**	**1,015.50**

(Source: District Revenue Office, Warangal)

4.14 Debt Waiver and Debt Relief Scheme

The Finance Minister, in his budget speech (2008-09) announced a Debt Waiver and Debt Relief Scheme for farmers. The scheme covers all direct agricultural loans extended to Small, Marginal farmers and Other farmers by scheduled commercial banks, RRB's and Co-operative banks.

A) Benefit to Small and Marginal farmers:
Cultivation of up to 2.00 hect. Completed waiver includes interest, complete waiver of restructured and rescheduled loans during 2004 and 2006 or on account of natural calamities.

B) Benefit to Other farmers:
Debt relief of 25% of the loan together with applicable interest of Rs,20,000-00, whichever is higher, as a one time settlement. Balance, together with the applicable interest shall be paid in a maximum of 3 instalments by the following dates.

1) 1st instalment – 30-09-2008
2) 2nd instalment – 31-03-2009
3) 3rd instalment – 30-06-2009

Investment Credit: It covers term loans for direct agriculture activities such as wells, pump sets, tractors, bullocks, plantations, horticulture, allied activities, diary, poultry, sheep rearing, fishery, piggery, bee keeping, green house and biogas.

The scheme will apply to loans distributed from 1-4-1997 to 31-03-2007. The lending institutions shall issue certificates to the effect of the waiver from 8-07-2008 onwards. It was programmed to organize RUNA VIMUKTHI MELA from 8-7-2008 to 15-7-2008 in Warangal district.

District level Monitoring Committee on Debt waved off Debt relief was conveyed on 19-06-2008 and on 27-06-2008. The particulars of Debt waiver and relief scheme of the district is:

Debt Waiver		Debt Relief		Total (Rs: in crores)	
No.of farmers	Amount	No.of farmers	Amount	No.of farmers	Amount
2,65,674	444.71	29,247	58.23	2,94,921	502.94

(Source: Lead Bank, Warangal)

Debt waiver and Debt Relief Scheme details of Warangal District. (2008)

Sl.No.	Name of the Bank	Debt Waiver		Debt Relief		**Total**	
		No.of farmers	Amount (crores)	No.of farmers	Amount (crores)	**No.of farmers**	**Amount (crores)**
1	Allahabad Bank	1,376	2.70	280	0.65	**1,656**	**3.35**
2	Andhra Bank	22,548	41.08	4,149	8.11	**26,697**	**49.19**
3	A.P.GrameenaVikas Bank	24,295	48.83	2,449	5.23	**26,744**	**54.06**
4	Bank of Baroda	5,008	10.79	1,941	3.89	**6,949**	**14.68**
5	Bank of India	129	0.23	22	0.05	**151**	**0.28**
6	Bank of Maharashtra	3	0.02	2	1.01	**5**	**0.03**
7	Canara Bank	4,299	7.61	1,130	2.32	**5,369**	**10.13**
8	Central Bank of India	3,416	6.76	587	2.77	**4,003**	**9.53**
9	Corporation Bank	2,165	4.81	339	0.60	**2,504**	**5.41**
10	ICICI Bank	124	0.06	11	0.02	**135**	**0.08**
11	Indian Bank	1,488	1.95	452	0.99	**1,940**	**2.94**
12	Indian Overseas Bank	9,324	10.63	1154	1.91	**10,478**	**12.54**
13	ING Vysya Bank Ltd.	876	1.68	160	0.33	**1,036**	**2.01**
14	Punjab National bank	792	1.50	107	0.20	**899**	**1.70**
15	State Bank of Hyderabad	31,724	43.50	3011	5.83	**34,735**	**49.33**
16	State Bank of India	29,545	58.76	6,601	12.17	**36,146**	**70.93**
17	Syndicate Bank	3,543	8.28	825	1.58	**4,368**	**9.86**
18	Union Bank of India	79	0.14	27	0.14	**106**	**0.28**
19	WDCCB	1,25,000	195.18	6000	11.43	**1,31,000**	**206.61**
	Total	**2,65,674**	**444.71**	**29,247**	**58.23**	**2,94,921**	**502.94**

(Source: Lead Bank, Warangal)

4.15 Different Policies and Schemes Introduced by the Government in the Agriculture Sector of Warangal District

Many Agriculture allied activities through the organization of innovative programmes have been introduced to increase the farm income in the areas of technology, extension, credit etc. This has resulted in the harvesting of even higher food grain production during past years.

Free Power Supply to Farmers

The farmers of Andhra Pradesh, as part of a promise made during the elections, have been provided free electricity for agricultural purposes. Power supply will be provided in shift systems on the basis of seven hours per day with some limitations based on the acreage of the cultivation. Some restrictions have also been laid on the crop pattern and the crops to be cultivated in that particular period (Rabi or Kharif). These uncollected power bills will be paid by the Government. Besides, since the power is provided free of cost, there would not be any cases of power pilferage. This meant that the State Government would also be paying for the loss incurred by AP Transco so far on account of unauthorized agricultural and one-bulb connections in cash. In the case of the agriculture sector, the commission endorsed the Government's free power scheme and the internal arrangements thereto. Accordingly, farmers cultivating dry lands and those possessing less then 2.5 acres of wet lands and using up to three pump sets did not have to pay anything whether or not they implemented demand-side management devices like capacitors. Corporate farmers and IT assessees will have to pay Re. 1 per unit with such devices and Rs. 2 per unit without them.

Jalayagnam

'Jalayagnam', as the word indicates, is a ritual for water conservation. It has been implemented by the Andhra Pradesh Government for the cultivating people of the state to bring 82 lakh acres under irrigation in five years. This project accords the highest priority for the development of irrigation infrastructure, particularly in backward and drought prone areas by taking up this program in a big way. Jalayagnam includes a number of irrigation projects that involve the construction of reservoirs and lift irrigation systems for lifting water from major rivers, particularly from the Godavari, to provide immediate irrigation benefits.

Jalayagnam programme aims to complete 32 major and 17 medium irrigation projects at a cost of Rs. 65,000 crores to provide irrigation to an extent of 71 lakh acres besides stabilizing an existing ayacut of 21.32 lakh acres. The programme also aims to provide drinking water to a population of 1.2 crore and to generate power to the tune of 2700 MW. Eight of these projects were to be completed before the kharif season of 2006.

The irrigation development was virtually stagnant for the last decade due to negligence towards the agriculture sector on which approximately 70% of the population depends. The fact that 26% of the population, mostly in rural areas, was living below the poverty line makes one wonder whether science and technology has made a significant difference to the quality of life in rural areas compared to towns and cities. In addition, the many farmer suicides in the state made the Government place more emphasis on and focus its efforts on the irrigation sector. Having recognized the importance of the agricultural sector for faster economic development, the government started concentrating efforts on its development and introduced several schemes for the benefit of farmers, one of which is the Jalayagnam project.

In the past 50 years, a total ayacut of 65 lakh acres was developed in the State. Jalayagnam is expected to double the area under irrigation. Jalayagnam in Andhra Pradesh constitutes a major component of the National Irrigation Mission's (NIM) target of bringing one crore hectares of land (2.5 crore acres) in the country under the plough.

The large allocation of funds for irrigation is carried out because the government considers irrigation an important growth engine. The most expensive of these projects are the lift irrigation projects intended to irrigate the dry Telangana region and supply water to the Anantapur-Ongole-Mehaboobnagar area which is "turning into a desert". With the land being 300 metres above water level, lift irrigation is seen to be the only way to take water to the dry region. Currently, 3,000 tmc of water from the Godavari is wasted into the sea as the land is elevated. For the first time in the post-independence era, the Congress government took up the task of linking the Godavari and the Krishna by constructing the Indira Sagar Project and the Rajiv Sagar project across the Godavari. By executing projects like Rajiv Sagar, Indira Sagar flood flow canal, SRSP Phase-II and so on, 21 lakh acres of parched land in Telangana region will be made fertile.

Strategies for Sustainable Agriculture

Based on the supposed irrigation potential during 2006-08 and the socio economic condition of the farmers of Warangal district, a production plan has been developed by the Agriculture Department so as to cover greater areas under cereals, millets, oil seeds and other dry crops. Care has been taken to reduce the areas that cultivate high risk crops like paddy, cotton, and chillies.

Mass Farmers' Orientation Training Programme

This programme is being implemented in the district over the past 10 years, in order to convert paddy crops irrigated by wells and bore wells into I.D crops in order to save power and water use and to increase the growth of pulses, oil seeds, maize and other commercial crops.

Adarsha Rythus

It is felt that the benefits of the Agriculture Department and allied activities programmes can effectively reach the farmer with an increase in personal contact and discussions. Therefore, in order to strengthen the extension mechanism, a novel approach to bridge the extension gap has been evolved through the support of 1219 Adarsha Rythus in the 1st spell and 1089 in the 2nd spell in the District, so that each Adarsha Rythu could cover 200-250 farm holdings.

For the successful implementation of Polambadi and to increase awareness levels among farmers with increased crop production, it has been decided to strengthen extension services in the District through the Adarsha Rythus mentioned above. The Adarsha Rythus will be positioned as facilitators at the village level for the purpose of generating more awareness among farmers with regard to advanced crop planning, reduction in the cost of cultivation, increase in crop production, live stock production, horticulture crops, micro irrigation practices, marketing issues, post harvest technology, credit related issues, irrigation and other allied Agricultural activities. Basic training and capacity enhancement training have been conducted for the 1st phase of Adarsha Rythus so far.

Soil testing

In order to optimize crop yields, the testing of soil is most important. Soil Testing is carried out by the Soil Testing Lab in the District. Soil samples are analysed for soil pH (Soil reaction) and Electricity Conductivity (Soluble salts and available Macro nutrients i;e.,N,P & K). The STL in Warangal is equipped with an automatic absorption spectrophotometer to help estimate the status of micro nutrients in the Soil, especially Fe, Cu, Zn and Mn. A total of 1267 Soil Testing kits were distributed and 310 Refills were provided to Warangal District, so as to deliver one each to every Grampanchayat. A training programme to help in explaining the use of these kits was already conducted at the Mandal level. Each kit will be able to analyse 25 soil samples each for the presence of pH, OC, N, P, K, S & Ca.

Seed

Seed plays a vital role in Agriculture production. Quality seeds suited to different Agro Climatic conditions and in sufficient quantity at affordable prices are required to raise productivity.

In Warangal District, about 70% of the farmers are dependant on public and private Seed suppliers viz., APSSDC, NSC, HACA and Co-operative Societies which are under the regulation of SA 1996 and Sco1983.

To ensure timely availability of good and quality seed, without much dependence on the above agencies, the Government has launched a scheme named Seed Village Programme to produce quality/ certified seed in the farmers' fields.

State Seed Multiplication Farm

The main object of the seed farm is the production of foundation seed in order to meet the production requirement for certified seed under the seed village programme. The required cultivation, certification, storage and processing charges will be met from the revolving funds that accrue out of the sale proceeds of foundation seed.

Seed Regulation

A total of 450 Seed samples have been collected and sent for analysis as against the annual target of 950 samples so far. It has also been proposed to dram and analyse service samples at 5 each in all Mandals on payment of Rs. 50 per sample.

Crop Diversification Programme

Indiscriminate cultivation of certain crops in marginal soils and cultivating without following seasonality and mono cropping often leads to the failure of crops, reduction in productivity and production and an increase in the incidence of pests and diseases. Therefore there is a need for the diversification of such crops and their replacement by crops like maize and jowar. Maize is ideal due to its adaptability and high productivity and jowar for having the highest tolerance to drought. Similarly, groundnut can be intercropped with red gram and castor can be grown in rotation with millets like jowar and bajra.

Fertilizers

The fertilizers required are provided to farmers on the basis of the crop pattern and the proportion of cropped areas. This is due to the fact that fertilizers also play a major role in the productivity levels of the crop cultivated. A target of 500 samples of fertilizers is allotted to Warangal District, out of which 82 samples have been drawn so far. The prorate shortfall in providing fertilizers is caused at times due to discrepancy in the arrival of stocks, preference and allotments made to Institutional agencies and immediate liquidation of stocks.

Reclamation of Problematic Soils

The main objective of this programme is to reclaim soil affected with alkalinity and to improve soil fertility, thereby increasing crop productivity. During the previous years in the district, the application of Gypsum resulted in increased production and productivity besides improving the quality of produce. Gypsum is supplied at a 50% subsidy for 2 hectares per farmer. So far, 687 MTs of Gypsum have been supplied to the Warangal District farmers.

Supply of Zinc Sulphate

The deficiency of Zinc has been noted to be greater than other micronutrient deficiencies in the district. To overcome the problem of zinc deficiency and to increase productivity, 31 MTs of $ZnSo_4$ at 50 % subsidy has been supplied to the farmers so far.

Organic Farming

An increase in awareness about and care towards the health ion in recent years has created a good demand for organic food products.

The scheme, "Establishment of Vermi Compost" units on subsidy has been taken up at 20 units per Mandal with the aim of promoting the usage of organic manure by farmers. So far, 85 units have been achieved in the district. Preference should be given for farmers under the certification of organic farming scheme for the organizing of Vermi compost units. The unit cost of each Vermi compost unit is Rs. 6,500. This is inclusive of the cost of a semi pucca shed at Rs. 4000 and the cost of 20 kgs of earth worms at Rs. 2,500. In addition, the subsidy on the total unit is 75% and this comes to Rs. 4,875 per unit. Training is provided to farmers for creating Vermi compost.

Vermi Hatcheries

This scheme has been enhanced to supply earthworms to farmers for creating a Vermi compost unit. The financial assistance comprises 75% of the capital cost, limited to Rs. 75,000 per unit.

Green Manure

The scheme is programmed to supply Green Manure seed to farmers in the district. The Agricultural Officer should conduct a germination test locally and proceeds for distribution only after ensuring good germination. Immediately on receiving report of any substandard supply of seed, further supply should be stopped and arrangements must be made for the replacement of seed besides ensuring the further action of intimating the District Joint Director of Agriculture.

Bio-Fertilizers

This scheme proposes to promote the usage of new species of Bio-fertilizers like Azospirillium and Azoctobacter by supplying them at 50% subsidy.

Pesticides

Warangal district has been given the target of drawing 315 samples. To ensure the quality of pesticides being sold in the market, the provisions of IA 1968, IR 1971 and IO of 1986 are being strictly implemented in the district. Out of the target, 103 samples have been drawn so far.

Polambadi

In the Polambadi programme, each grampanchayath is being facilitated with farmers' field schools and these have also been organized as part of various other projects such as the Andhra Pradesh community based Tank Management Project, National Food Security Project, ICDP (cotton), Isopom (Maize, Pulses and Oil seeds) Work Plan (Rice), training in organic farming and training in seed production. A total number of 1018 Polambadis have been allotted to the district so far. Out of these, 518 are already functioning and the remaining have yet to be started.

Farm Mechanisation

Farm mechanisation involves the judicious application of inputs through the use of Agriculture machinery/ equipment, viz, hand sprayers, bullock drawn equipment and power driven machines in order to perform the various operations required for crop production activities. Farm mechanisation provides the technology required to facilitate Agriculture through the efficient utilization of inputs and timelines of operations, besides reducing drudgery. During the current year, a financial target of Rs. 500 lakhs has been allocated. Out of this, 186 lakhs has been spent to supply5616 Nos. of various implements.

National Watershed Development Project for Rain Fed Area (NWDPRA)

The NWDPRA scheme has been restructured and its guidelines have been named as "WARASA JANA SABHA GEETA". To arrest the growth of deserts under natural resource management, Agro Forestry, Horticulture, Pasture development, Soil and moisture conservation measures have been taken up in the watershed area. Under the farm production system, diversification of production system testing and demonstration of new technologies, adoption of proven technologies such as organic farming, use of bio fertilizer INM and IPM development of Micro irrigation systems have been taken up.

As part of the IX Financial plan, 5 watersheds are being organized in the villages of Palakurthy Mandal in the District with the financial assistance of Rs. 6 lakhs.

Credit

In order to facilitate the grant of Agriculture loans for tenant, Small-Marginal farmers and other farmers from banks, the scheme has been programmed to distribute crop loans to a tune of approximately Rs. 760 crores in the district during 2008-09. Out of this amount, Rs. 117 crores have so far been extended as loans.

Loan Waiver Scheme

The Government has introduced the "RUNAVIMUKTHA MELA" scheme to waive off crop loans taken by farmers from 31-12-2007 to 29-02-2008. Accordingly, the Mandal level runa vimuktha melas are also being conducted. So far, 401.6 crores have been waived off, covering 177576 farmers in 37 Mandals out of an estimated target of Rs. 502.94 crores.

Commodity Interest Groups: (CIGs)

Carrying out group extension methodology involves organizing training demonstrations, farmers' field schools, seed production etc. These groups are formed into federations and develop linkages to provide marketing information and support etc. They formulate methods to ensure optimal resource use in plants for production and marketing. These groups function to co-ordinate with financial institutions, research institutions and the corporate sector and strive to establish long term relationships for the benefits of its members.

Agricultural Technology Management Agency (ATMA)

This scheme is proposed to be funded on 90:10 patterns between the government of India and the state. It is a society consisting of key stake holders involved in agricultural activities for sustainable agriculture development in the district. It is a focal point for integrating research and extension activities, thereby decentralizing the day to day management of the public agricultural technologies system.

The scheme shall be implemented through autonomous district level institutions to be established by the state in the form ATMA. The Commissioner and Director of Agriculture, Hyderabad has been identified as the Nodal Officer at the state level.

Implementation of National Food Security Mission (NFSM):

The district level committee for the effective implementation and monitoring of the scheme, Government of A.P has constituted the State Food Security Mission: Executive Committee (SFSMEC) and District Food Security Mission: Extension Committee (DFSMEC).

The DFSMEC has been constituted under the Chairmanship of the District Collector, Warangal with representatives from the Line Departments, nominated progressive farmers and representatives from self help groups of the farmers/ CIG's representatives from reputed NGOs.

4.16 Field Study

4.16.1. Reasons for Selecting Warangal District for Survey

The primary reason for selecting Warangal district for the purpose of study is that, the district basically has an agrarian economy status. The major pre-occupation of people in the district is agriculture and allied activities.

The second reason is the farmer suicides in Andhra Pradesh that have garnered country wide attention and among the total number of these suicides, Warangal district tops the list by displaying the maximum number of suicides in the recent era. Though farmer suicides have occurred in almost all districts of Andhra Pradesh, they were very great in number in the district of Warangal, which displayed the highest percentage (nearly 80%) of the total suicides recorded in the district (E. Revathi).

The third reason is that, although the recent policies of the government have ensured many supportive measures to farmers to avoid suicides and to eradicate their dismal conditions and despite a slight fall in the rate of farmer suicides, a considerable rate of farmer suicides still continues to exist in the district. According to the District Revenue Department Statistics during the period of 1998-2007, the total number of farmer suicides in the district was recorded as 677. If one considers only the recent 5 year period between 2003-2007, when farmer suicides were recorded as totalling 282 in number, it can be considered as the highest farmer suicide rate in the state as well as in the country.

This dismal condition of farmers in the district has attracted my attention to the study of acute and unprecedented conditions in the district, in the past as well as in recent years.

As the district is not abundant with irrigational resources, it is mostly dependent on rainfall for cultivation. Due to a continuous drought, the district has undergone severe crop failures and this has resulted in an agrarian distress in the district. This has consequently resulted in the highest rate of farmer suicides in the country.

To study the reasons for the above mentioned conditions in the district and to suggest some measures to overcome such situations, I have conducted a field survey in Warangal district. Choosing three different mandals like Ghanpur (Stn.), Dharmasagar and Atmakur of the Warangal district as samples, I have attempted to examine the conditions of the families of deceased farmers and have tried to isolate the reasons behind their suicides. My main aim was to discover why, though there have been recent policies which are projected as being farmer friendly in nature, farmer suicides are yet to be eradicated to a large extent. To this end I have selected 10 households from three different mandals. In addition, my study will also focus on those farmer suicides that have been recorded in the District Revenue Office records on the basis of the three member committee constituted by the District Collector. Priority has been given to those farmer suicides which took place during the period between the years 2003-2007.

To seek out the reasons behind the distress, I have tried to gather and analyse data on different aspects of the living conditions and agricultural practices of those deceased farmers considered as part of the sample. They are as follows:

A) Family Composition:

The particulars of the family composition of the 10 selected farmers have been shown in the following table.

Particulars of Family composition of selected farmers

Sl.No	Category	Total	Average
1	Male	7	0.7
2	Female	6	0.6
3	Children	21	2.1
4	**Total**	**34**	**3.4**

Source: Field study / computed.

As can be seen from the above table, the average size of the family in the district is 4. The number of adult males and females in a family on an average is works out to 1 and 1 respectively, which is in tune with the sex ratio of the district as a whole (see general statistics of the district in Chapter IV). In other words, it can be said that the supply of labour to agriculture is 2 from each of the selected families.

The study shows that the selected families possess more assets in value than their liabilities. But still, they are not able to come out of their miseries since most of these assets are non-liquid assets such as land and depreciable durable goods. Moreover, their agricultural income happens to be low and uncertain.

i) Assets:

a) Land particulars:

The details of the area of land owned by the selected farmers with all the attendant classifications such as land used for cultivation, land leased in, etc have been displayed in the following table.

Particulars of Land Owned by the Selected Farmers (Acres)

S.No	Category	Irrigated	Dry(acres)
1	Land used for cultivation	--	45
2	Average land owned per farmer	--	4.5
3	Total land leased in	--	10
4	Total land leased out	--	--
5	Total land area of the farmer	--	35
6	Average land/farmer	--	3.5

Source: Field study/ Computed.

Total land acres owned by the selected farmers do not include irrigated land as the mandals selected for the study hardly have any irrigational resources. The dry lands of these farmers constitute 35 acres, which includes both land owned by selected farmers and the land leased in by them. The total land under cultivation includes both leased in and owned lands. Leased lands constitute a small extent of the total land area of the farmers, that is, 10 acres. Dry lands constitute 45 acres and the average land owned per farmer is 3.5 acres.

Category of selected farmers: The selected farmers have been categorized into small, medium and large. The categorization has been carried out on the basis of the method used by the department of economics and statistics for preparing the Agricultural census. As per this method, small farmers are those that own land area that measures between one acre and two acres. The farmers who own two to four acres have been categorized as medium farmers and the farmers who own 5 and more than 5 acres have been categorized as large farmers.

Among the selected farmers, small and medium farmers total seven in number. The remaining three farmers are large farmers.

b) Value of Land:

The land in Warangal is fertile and suitable for agriculture. Because of erratic and unsuitable cultivation practices and the practice of following single crop cultivation without a shift in the crop pattern however, soil degradation and water level decline has occurred. Again, because of the drought and climatic variation in the district, people do not purchase land in the district, even though the farmers are willing to sell. The value of land has thus become very low and now in many parts of the district, the government has fixed the price for the land. The details regarding the land value of the selected farmers is shown in the following table.

Particulars of the Value of Land Owned by Selected Farmers (in Rs.)

S.No	Category	Total(Rs)	Average value/household(Rs)	Average value/acre(Rs)
1	Irrigated	--	--	--
2	Dry	20,15,000	2,01,500	44,777.77
	Total	**20,15,000**	**2,01,500**	**44,777.77**

Source: Field study/ Computed.

Land is considered as an important asset and therefore its value has much importance in the category of assets. The average value of irrigated and dry land per acre works out to be Rs 44,777.77. The demand for land has been falling continuously in the district and it has forced the government to fix price limits for land. The value of one acre of dry land in the district is now approximately Rs 30,000 to 70,000.

i) Livestock:

Farmers in the Warangal district substantiate the view that livestock is an important sub-sector of agriculture. Livestock has emerged as an important supplementary source of income for farmers in the district. But rearing livestock had become a problem in the district due to the destruction and decline of the fodder base. Hence the livestock of the farmers selected for this study has, with the exception of bullocks used in plugging the fields, been sold off due to the bad grazing conditions brought on by continuous drought. So the livestock of the selected farmers can be treated as comprising a negligible ratio. The total value of the livestock of the selected farmers can be estimated at Rs 1, 20,000 and the average value of livestock per farmer is 12,000.

ii) Farm Machinery and Implements:

It can be observed that the farmers selected from the district are those farmers who use age old practices of cultivation. They still use ploughs and other traditional implements for cultivation instead of new, technologically sophisticated implements. This may be on account of the farmers' small land holdings or perhaps because they are not able to afford such expensive implements. Though these implements cut the cost of cultivation, these selected farmers are not able to afford those farm machinery and implements like tractors, sprinklers, sprayers and motors, even though some of them are available at subsidized prices. If they need such implements, they generally prefer to hire them instead of owning them. Thus the total value of the farm machinery and implements of the selected farmers is Rs. 40,500 and the average value per farmer is Rs.4, 500, which can be considered as a negligible value.

iii) Irrigation:

Irrigation is the most important input for agriculture. Especially in Andhra Pradesh, Agriculture is heavily dependent on the monsoons. In such a case, alternative ways for the provision of water are essential. Warangal lags behind in the provision of irrigational facilities. Even though recent policies have paved the way for new irrigational projects, most of them are yet to be finished while some of them remain stagnant at the foundational level itself. The value of the selected farmers' irrigational facilities can be considered as a minute proportion of the farmers' assets as the cultivation is highly dependent on rainfall. The total value of the irrigation facilities is Rs. 53,000. Though a couple of farmers own wells, these are totally dried up. Some farmers tried to dig bore wells and such efforts ended futilely. These attempts trapped the farmers in debts. as their hope of finding water and alternative irrigational sources for cultivation made them invest a lot of money in digging bore wells in vain.

Source of Irrigation

S.No	Category	No. of Beneficiaries	Area Irrigated	Value
1	Canal	-	-	-
2	Tank	-	-	-
3	Well	3	11	43,000
4	Bore well	1	2	10,000
5	Rain fed	6	32	-
	Total	**10**	**45**	**53,000**

Source: Field study/Computed

It is a fact that irrigation facilities are highly lacking and that minor irrigation projects, which will be more useful to farmers, are lacking in the areas that need them.

ii) Buildings:

Another important component of the assets is the buildings. Among the selected farmers, each one possessed a house of his own. These buildings vary from katcha houses to pukka houses and the total value of the buildings can be aggregated as Rs. 5, 90,000 and the average value per farmer can be estimated as 59,000.

iii) Savings, Deposits and other Assets:

Owing to the continuous drought prevailing in the district, the savings, deposits and liquid amounts of the selected farmers also constitute a minor fraction. To some extent we can say that none of the farmers have any savings at all. When it comes to these farmers' liquid assets, the amount they hold is just sufficient to meet the day to day needs of their families. Since they have faced continuous losses in agriculture, all the selected farmers possess a minute ratio of savings. The other assets they posses include gold, silver and some house hold goods like T.V, radio etc. Moreover, a portion of these selected farmers' gold and other assets are with money lenders or banks, handed over by them as a mortgage for credits and loans. The total value of the Deposits and Assets of the selected farmers is computed as Rs.1, 25,500 and the average value per farmer is Rs.12, 500.

Finally, we can arrive at an overall view of the assets as a whole, of the selected deceased farmers, that have been selected as samples for the study. Since we are considering the asset-liability ratio as an important indicator, the total asset value has much importance.

Total Value of Assets (in Rs.)

S.No	Category	Total Value	Avg. Value/ Farmer
1	Land	20,15,000	2,01,5000
2	Livestock	1,20,000	12,000
3	Farm Machinery & Implements.	40,500	4,500
4	Irrigation	53,000	5,300
5	Buildings	5,90,000	59,000
6	Deposits & Other Assets	1,25,500	12,550
	Total	**29,34,000**	**2,93,400**

Source: Field study/ computed.

The total value of the assets computed for the selected farmers is Rs.29, 34,000. The average value per household is Rs.2, 93,400. Land is the major asset of farmers in the district. It constitutes 72% of the total value of assets, followed by buildings and then deposits and other assets.

B) Liabilities:

As regards liabilities, all the sample farmers have debts. We cannot isolate a single farmer without a debt or loan in the district. Most of the farmers however, tend to borrow money from money lenders rather than banks or co-operatives. This is primarily because the process getting credits or loans from banks and co-operatives involves many sureties and lengthy paper work. In addition, there is a great delay in receiving the loan amount from banks and co-operatives. It has been observed that most farmers have both agricultural and non-agricultural loans. The loans taken for agricultural purposes however, are greater in number in the case of the selected farmers. These agricultural purposes include inputs purchase, bore well digging, well repairs, motor repairs etc.

The non-agricultural purposes include expenditure for marriages, health needs, and household expenditures.

Source wise and Purpose wise Loans

S.No.	Source	Agricultural Purpose	Non-agricultural purpose	Total
1	Co-operative banks	5,80,000	-	5,80,000
2	Money lenders	17,20,000	2,00,000	19,20,000
3	Private banks	1,00,000	50,000	1,50,000
	Total	**24,00,000**	**2,50,000**	**26,50,000**

Source: Field study/ Computed.

The total loan amount of the selected farmers is computed as Rs.26, 50,000. This amount includes loans from both organized and unorganized sources like banks, co-operatives and money lenders.

The rate of interest varies from 0.60 % to 3%. The average liability of each selected farmer constitutes Rs.2, 65,000.

Extent of Liability among Selected Households

Loan Amount (Rs.)	No. of Households
5,00,000 and above	-
4,99,999 – 4,00,000	2
3,99,999 – 3,00,000	1
2,99,999 – 2,00,000	6
1,99,999 – 1,00,000	1
9,99,999 and below	-

Source: Field Study/ Computed

Asset-Liability Ratio:

The asset-liability ratio is considered as an important indicator for assessing the debt repaying capacity of the farmers. The ratio calculated on the basis of the data collected is 7:3. As stated earlier, the asset value surpasses the liability value. But the non-liquidity of the asset forms indicates that they are least helpful in improving the repayment capacity of the farmers.

C) Cropping Pattern:

The cropping pattern of the selected farmers in the district tends mostly towards the cultivation of paddy or cotton. The cropping pattern in the district is different on account of the expectations of returns and the suitable availability of resources for crops like cotton and paddy. Though the district cropping pattern comprises varied crops like bajra, jowar, chillies, tobacco, cotton, paddy etc., the selected farmers cultivated only cotton and paddy.

Particulars of the Cropping Pattern of Selected Farmers

S.No	Crop	Acres	% in total cultivated area	Avg./ Farmer
1	Cotton	32	69.4	3.2
2	Paddy	13	30.6	1.3
	Total	**45**	**100**	**4.5**

Source: Field study/ Computed

We can see from the above table that cotton plays a vital role in the cropping pattern of the selected farmers. In addition, the crop failure which resulted in the state wide suicides is considered to be mainly due to the failure of cotton crops. Even in the case of the selected farmers, the crop pattern clearly explains the high rate of cotton cultivation in the district.

C) Cost of Production of Selected Crops:

The cost of production in the agricultural crops being discussed has been classified under two different categories, material charges and labour charges. The material costs include the cost of purchasing farm implements, manures, pesticides and plant protection materials. With respect to farm implements like sickles and ploughs, the farmer may not require to buy them every year. Still, in the case of the selected farmers, the material cost is higher than the labour cost. This may be because the farmers would have been forced to use more protective materials such as chemicals, fertilizers etc. The total material cost incurred by the selected farmers is Rs.6, 00,900.

Material Cost per Acre of the Selected Farmers

S.No	Crop	Total Material Cost(Rs.)	Avg./ Crop acre
1	Cotton	4,83,500	15109.37
2	Paddy	1,17,400	9030.76
	Total	**6,00,900**	**24140.13**

Source: Field study/ Computed.

The total labour cost of the selected farmers in the district is computed as Rs.86, 550. This is very less compared to that of the material costs incurred by the selected farmers in the district. This shows the clear variation in the cost of cultivation of crops in the district. The labour cost incurred by the selected farmers for cotton crops is Rs. 59,560. For the Paddy crops, the selected farmers incurred Rs. 26,990. In the case of the paddy crop however, the average labour cost per crop per acre is higher in comparison to that of cotton owing to the limited rate of acres where paddy has been cultivated.

Labour Cost per Acre of the Selected Farmers

S.No	Crop	Total Labour Cost (Rs.)	Avg./Crop Acre
1	Cotton	59,560	1861.25
2	Paddy	26,990	2076.15
	Total	**86,550**	**3937.40**

Source: Field study/ Computed.

The following table summarizes the material cost and labour cost and gives us information about the total cost of production incurred by the selected farmers in the district.

Total Cost of Production of Selected Crops by the Selected Farmers (in Rs.)

S.No.	Crop	Material Cost (Rs.)	Labour Cost (Rs.)	Total Cost (Rs.)
1	Cotton	4,83,500	59,560	5,43,060
2	Paddy	1,17,400	26,990	1,44,390
	Total	**6,00,900**	**86,550**	**6,87,450**

Source: Field study/ Computed.

The total cost of production incurred by the selected farmers is Rs.6, 87,450. This comprises both material costs and labour costs incurred during the cultivation of the selected crops by the farmers. As the cultivation of cotton is more than other crops, the total cost incurred for it is greater in the district. The table shows that the material costs are higher than that of the labour costs and this implies that the greatest usage of protective costs in production is due to the increased usage of pesticides, fertilizers etc.

D) Yield of Crops:

Yield per Acre is an important indicator of agricultural productivity. The following table illustrates the productivity of selected crops cultivated by the selected farmers of the district, in material as well as value terms.

Yield/Acre and Value

Sl.	Crop	Total Yield(Kg)	Total value (Rs.)	Yield/Acre (Kgs)	Value/ Acre (Rs)
1	Cotton	8,800	1,33,100	275	4159.375
2	Paddy	10920	1,16,700	840	8976.9231

Source: Field study/Computed.

Though the paddy yield shows better productivity levels, the higher returns forced farmers to cultivate more of cotton. Thus the cultivation of cotton took place in 32 acres, whereas paddy was taken up only in 13 acres. But for these farmers, there were no proper irrigational sources and this, along with continuous drought in the district, resulted in a crop failure. The farmers did not even get their investment back and this in turn has trapped them into debts due to a very low yield per acre. In a period of good crop, one acre of land will produce 6 to 8 quintals of cotton, whereas one acre of paddy may produce 25 to 40 bags depending on the sources provided. But owing to continuous drought for several years, only 2 to 3 quintals of cotton were produced per acre and 10 to 20 bags of paddy were produced per acre. It can thus be inferred that the farmers didn't even get their investment back.

E) Net Returns:

This is one of the most important indicators in the study and the following table helps give an idea about whether or not agriculture is remunerative.

Particulars of the Returns on Production of the Selected Farmers (in Rs.)

S.No	Crop	Total Revenue	Total cost of Prodn.	Profit	Loss
1	Cotton	1,20,100	5,43,060	-	4,22,960
2	Paddy	81,500	1,44,390	-	62,890
	Total	**2,01,600**	**6,87,450**	**-**	**4,85,850**

Source: Field study/ Computed.

The primary data that I have computed on the basis of information provided by the selected farmers in the district, leads us to the conclusion that agricultural is a losing proposition for it is not remunerative. The cost of production of the crops has been observed to be more than the returns. The losses incurred in the returns on the crop are beyond repair. The farmers are generating heavy losses due to a decline in productivity, pest attacks and fall in prices.

The total cost of production of the selected farmers is Rs. 6, 87,450 and the total returns they received on the yield is Rs.2,01,600. Instead of profits therefore, they have reaped huge losses amounting to Rs. 4, 85,850. These losses can be said to have a greater inverse proportion to the cost of production. Besides, this condition prevailed for a continuous period due to incessant drought in the state. This trapped farmers in debts and they were left with no other source with which to repay these debts. This situation left the farmers without any alternative solution, as their only knowledge is of agriculture and they believe strongly in their age old occupation. Hoping for a good crop every year, these farmers invested new amounts borrowed through different sources on their crops. However, their fate was the same every year as they experienced only crop failures and thus all their efforts had been in vain.

As a whole, the cost of production has become higher on account of high inflation and changing economic conditions. Therefore both material as well as labour costs have increased and made it a burden for farmers to cultivate their lands. Moreover, the policies of liberalization have further added to the farmers' situation and made it worse. In such a condition, farmers are left without any means of survival and the pressure of their socio-economic and psychological conditions thus forced many of them to take their own lives.

Chapter Five

Conclusions and Summary

Identification of the Causes for the Distress

1) Micro Level

The present situation of agriculture in the agrarian district of Warangal is such that it is unable to provide decent living conditions for the families that are primarily dependent on it. These families once lived comfortably with the same resources of land irrespective of whether they were large or small farmers.

With respect to the quality of inputs, farmers used to take seeds from different sources like Government seed corporations and known persons. The quality of seeds however, was not good and this even resulted in crop failures due to fraudulent dealings with some private seed vendors. Some private seed vendors also created an artificial seed scarcity in order to increase the prices of seeds. However, with regulations and raids carried out by the Agriculture department and the Intelligence department, this problem was solved to a certain extent.

A majority of the farmers in the district are unaware of the programmes undertaken by the Agriculture department of the District. One reason for this might be the lack of proper information provided to farmers by the department. The other is the high levels of illiteracy among farmers of the district, which contributes to their lack of awareness about such developmental activities and new innovative technologies in the field of agriculture etc.

Loans and funds are always very important in the agriculture sector. It has been noticed that when the agriculture situations were better, banks were financing as much as the farmers' requirements but during the distress situation, most banks were not willing to provide agricultural loans to farmers as a part of their risk aversion. The bureaucratic delay in providing loans forces farmers to borrow from money lenders, friends etc., at a higher rate of interest.

Lack of technical guidance, pest and disease attacks, droughts, unseasonal rainfalls and floods are some of the agricultural problems faced by the Warangal district.

There are no significant problems in the marketing aspects of agricultural products because the district has been endowed with good markets which have a good name in the world. The Enumamula Market Yard is regarded as one of the largest market yards in Asia and is located in the District. According to farmers however, the problems faced during marketing include the lack of minimum support prices to the production, unremunerative prices to the yield when compared to the cost of production, a negligible rate of middlemen frauds and the competition laid by agricultural imports and the high transport charges of produced goods which need to be transported from villages to the markets.

2) Macro level

The major causes for agricultural distress at the macro level are regarding:

a) Technology,
b) Natural Resources and
c) Capital Resources.

a) Technology:

i) Rain fed Bias:

There is a demand for a high level of irrigational facilities in Warangal district because cultivation here is mostly dependent upon the monsoons and the major proportion of the soil types are dry areas which, due the low availability of moisture due to temperature situations, also form part of this demand for irrigational facilities. So far, though the policies of the government have stressed on the importance of irrigational facilities, they are yet to ensure the adequate provision of these facilities. This is resulting in crop failures and losses in crop production.

ii) Agro Chemical use:

The globalization era has led to the commercialization of the agricultural sector and has resulted in declining labour productivity and driven the farmers to carry out intensive use of fertilizers, pesticides and herbicides. There is a decline in input- output ratios, an increase in micro nutrient deficiency, a decline in soil quality and a rise in ground water pollution due to overuse or an unbalanced use of fertilizers. The unbalanced use of pesticides has fallen slightly due to the development of improved pest resistant varieties of pesticide. The use of herbicides is on the rise and is likely to increase further in the future. This forms a part of the technology used in farming. The cost of these chemicals is higher than their benefits, and once we consider environmental costs, it will be much more. Pesticides can affect human health through the residues they leave behind in food products.

iii) Total Factor Productivity (TFP):

Technology that heralded the process of the Green Revolution during the 70's and 80's started showing signs of deceleration during the 1990's.Technological stagnation is supported by the increasing evidence available about stagnating levels of productivity and crop growth even in many areas with potential. The good growth rates registered by the agricultural sector in the 1970's and 80's, showed deceleration from the beginning of 1990's. Non price prerequisite for sustaining the agricultural production namely, the technology receded in the later periods, and it's bearing on the profitability of crop is reflected in the output-input ratios of many crops. The surplus production of rice and wheat could not be exported profitably as the ruling prices in the international markets remained far below the cost of procurement. Consequently, with mounting stocks, the prices of the commodities in the domestic market fell far below the cost of production. The benefit-cost ratios for a majority of crops in many states of the country declined as a result of the decline in TFP. Decline in TFP and output-input ratios clearly substantiate that the returns to investment in agriculture seem to be declining. Farmers may have been able to maintain the yields through application of higher amount of non-land input, which are very costly, which again means a declining trend in TFP and profitability in farming.

iv) Mechanisation:

Mechanisation is complimentary to the modern varieties of technology to realize the full production potential. The share of machine-labour in the total cost of production has risen significantly whereas the share of human labour as well as animal labour has sharply declined. Of all the modern agricultural technologies introduced, mechanization has been widely discussed. It has been blamed for exacerbating rural unemployment. This drawback of mechanization necessitates the proper selection, utilization and management of farm power resources. Since efficiency demand and export demand are crucial, it is advisable to continue promoting mechanization in agriculture, provided concurrent programs and policies which could offer alternative employment for the poor in rural non-farm sector are designed.

b) Natural Resources

i) Land Degradation:

Land degradation is another serious problem that causes a decline in crop production. Sustainable management of resources is beset with the quantitative and qualitative problems of land. The quantitative dimension relates to issues like increasing human and animal population, pressure on land, changes in the land use patterns and erosion and loss of top soil. The qualitative dimension relates to the loss of nutrients and pollution of soil environment by agriculture and non- agricultural activities. Small and marginal farmers who cultivate annual crops are particularly affected by this land degradation. The economic and environmental impact of this soil degradation will be tremendous assuming that the current rates of soil degradation will continue unabated.

ii) Marginalization of Holdings:

To increase productivity, employment and income of poor segments of agricultural operations is considered to be a central issue in agricultural development. Small farmers constitute a sizeable portion among the rural poor in developing countries. During 1971-72 to 1991-92, it has been noticed that there was an intensification of the marginalization process. During 1971-72, a share of 33.97% in the total land emerged as a major sized group of small farmers and this fragmentation increased further in the next two decades. When the liberalization era began in the country, over 40% of the rural households were landless and the remaining 60% owned land holdings. Out of this 60%, over 96% of the households were near landless, but over two thirds of the owned land belonged to marginal, small and semi-medium sized groups. The large farmers constituted 1% of the total land owners in 1990-91, but owned nearly 13.83% of the total land. This trend of marginalization may continue in the future but at a slower magnitude (Ramsay, 2004).

iii) Water Demand:

There is tremendous pressure on both quantitative and qualitative dimensions of water resources in the country. This is reflected in the sharp decline in the per-capita availability of water. Increased competition as a result of demand for water is clearly highlighted by the water-related tensions and conflicts between states and rural and urban areas within the states. Agriculture is not only the single largest consumer of water, but is also increasingly becoming a cause of water quality deterioration. Production of modern crop varieties demands a larger supply of water. Scarce and viable water supply plays an important role in water resource management which should embrace the slogan 'Equity and Efficiency', since the distribution of water resources across different regions are highly uneven. Inequalities in access to irrigation water across farms within an area also exist.

c) Capital Resources

i) Agricultural Investment:

Agricultural development relies heavily on government finance due to the presence of externalities, high risks and inadequacies in agricultural credit and infrastructure. In recent years, the government expenditure on agriculture to total expenditure has not increased, but has declined. At present, the agricultural investment share of the government in the Tenth Plan was only 5.2%, which has fallen from a 14.7% share in the Eight Plan. Growth in agriculture in the early years was mainly due to the planned investment made in the agricultural sector through the expansion of irrigation facilities and the spread of fertilizer and credit outlets coupled with supportive price policies. Some of the empirical studies have shown that the instability in government expenditure is detrimental to agricultural growth. There should be supportive, adequate public outlay which functions as an important stimulus to agricultural growth in an economy.

ii) Capital Formation:

Without adequate investment, agriculture cannot make a substantial contribution to the economic development of a country. The share of agriculture in the GDP of the economy has been declining over the years and one reason for this is the declining investments by both public and private sectors and the resultant slow growth of capital formation. The public and private investments are complimentary rather than substitutes. Studies have found that the reduction in public farm investments depressed private investments by nearly 15% in India. But, in addition, profitability of production and the availability of formal credit are also determinants. The problems of the Indian rural economy can be epitomized in the words, "Indian farmer is born into debt, lives in debt and dies in debt". This has become apt in the case of farmers in Andhra Pradesh, especially in Warangal district. Right now, the crisis-ridden farmers in the district need essential support from the part of the state. Now the government will have to shift emphasis from direct interventions in the rural economy to undertaking supporting tasks. The government decision to hold back loan recoveries of farmers will not help them in any real sense. In this present juncture, what is badly needed is strong government support and policies that ensure the promotion of growth and benefits for the farmers without bureaucratic delay.

Summary

Agriculture is the soul of the Andhra Pradesh economy. This study has been conducted to analyse the agricultural problems plaguing the state of Andhra Pradesh. The study has been pursued with the objective of analysing the income, output and productivity conditions of the agricultural sector in Andhra Pradesh. It also attempts to probe into the nature of the distressful conditions prevailing in Andhra Pradesh's agriculture sector. The study also seeks to identify the causes for the crisis in the sector that has lead even to farmer suicides. Finally, an attempt has been made to suggest suitable and feasible measures to ease the prevailing conditions.

A case study has been conducted in Warangal, a predominantly agrarian district, in the state of Andhra Pradesh. From the Mandals of Ghanpur, Dharmasagar and Atmakur, ten farmers were chosen. The sources of data consist of both primary and secondary sources. The conclusions drawn from the study have been substantiated with the analysed and computed data.

The introductory chapter examines the dimensions of the problem with which the study deals. The second chapter reviews the available literature on general agricultural problems plaguing various parts of the country. The chapters that follow the introduction deal with the agrarian aspects of Andhra Pradesh in general and then of Warangal in particular. The final chapter summarizes the study and concludes with a brief description of possible remedial measures.

Regular crop failure and a lack of remunerative price for the produce are the main causes for the pathetic condition of the farming community in the state. The cost of cultivation has increased substantially over the last few years owing to a cutback in various input subsidies provided by the government. While perpetual subsidization of agrarian inputs cannot be a long-term measure for sustainable development, what has caused an enormous blow to the farm sector is the unexpected and sudden removal of some of the input subsidies.

The debt trap is a major aspect of the distressing conditions experienced by the agricultural sector in the state. Lack of easy access to institutional credit is now a key problem for the farmers. Crop failures have destroyed the repaying capacity of the farmers. The unpreparedness of the institutional credit sources to confront the unexpected problems of farmers has become a bottleneck that prevents an easy escape from the present agrarian crisis.

While self help groups and other alternative sources of funding that exist in the economy have a role to play, the scale of credit that they offer cannot be sufficient to meet the rising costs of agricultural inputs. However, the real challenges lie in providing for a sustainable and reasonable livelihood proposition for farmers. Till this happens, measures such as the provision of increased credit and moratoria on debt recovery are merely efforts to postpone impending doom.

The expansion of India's agricultural exports are being hindered by high import tariffs imposed by developed countries, high cost of domestic production of agricultural commodities, low competitive efficiency of Indian commodities, low agricultural export subsidies and the highly volatile international prices of agricultural commodities. The imports of agricultural goods have increased tremendously due to the lifting of the ban on the import of agricultural products, removal of quantitative restrictions on imports and the low level of import tariffs imposed by the government. The increased imports of the commodities have not only reduced the demand for domestic product but have also crashed their prices drastically.

The direct and indirect impact of liberalization policies on agriculture and the withdrawal of all forms of government support have exposed those who are dependent on agriculture to the fluctuations of the market. The traditional village support networks have also disintegrated under the pressure generated by the new market oriented dispensation. The prevailing patterns of declining crop yields, deceleration in the rate of growth of output and shift in the cropping pattern in favour of high-risk commercial crops are pushing the farming community into a desperate predicament. The importance of institutional credit availability has to be seen against this background.

The occupational structure of Warangal district is predominantly agrarian in nature. As is the case with the state of Andhra Pradesh as a whole, Warangal district too has a high rate of suicides. Socio- cultural, economic and other institutional factors play a crucial role in the high rate of suicides, which are often explained as being related to alcoholism, unemployment, poverty, frustration and depression. This agrarian district also has a high incidence of farmer suicides due to crop failures, bankruptcies and other financial liabilities.

Conclusions

The agrarian crisis in the economy is not of sudden origin. What is seen today is merely the re-emergence of problems after periods of temporary relief. With the decline of the positive effects of the green revolution, it is clear that temporary solutions cannot provide the way out for Indian agriculture. More importantly, the dawn of a liberalized regime, which happened without adequate internal reforms that could have corrected the course of Indian agriculture and made it competitive, remunerative and sustainable, has rendered temporary solutions rather irrelevant and in some cases, even unthinkable.

Decline in land and factor productivity seem to indicate a reversal of fortunes in the post green revolution agrarian sector in India. There is hardly any growth in real income generation across the sector. The contribution of agriculture to National Income has also declined. One needs to record that there has been a steady decline in public investment in agricultural infrastructure over the years.

The problems of the farmers in the state of Andhra Pradesh are not confined to any particular crop or region. Fall in production, declining productivity, fall in prices, increasing cost of production, mounting debts and increasing difficulty in meeting credit requirements seem to be common aspects of the problems faced by the farming community across the state. While economic reforms should have led to greater competitiveness and more efficient use of available resources, an unprepared agricultural sector seems to have only reeled under the pressures of a globalised market. A strategy of planned development has often appeared to have worked in the past and hence, the sudden shift made by the state from its role as initiator of development to that of facilitator of development, has apparently adversely affected the farming community.

However, it is clear that when the economic reforms were initiated with much excitement, the decision makers did not help much in equipping the farming community in meeting the requirements of a market-oriented system. Hence, it is not surprising that farmers across the country and in Andhra Pradesh, in particular, feel that their interests have not been taken care of. Absence of any safety-net programmes, which the states of developed market-economies in the west have actively implemented, meant that there was nothing to save the farmer from the economic repercussions of the vagaries of nature and the fluctuations in market conditions.

With no effective crop insurance systems, there is hardly any hope for a farmer who is experiencing drought. Variations in the amount of rainfall received therefore severely affects the farmers. Variation in the rainfall received can be attributed to the changing patterns of climate due to global warming. The insufficient irrigational facilities have meant that compensating for a shortage in rainfall has become almost impossible.

Literature on investment in agriculture often points to the complementary nature of public and private investment. Hence even theoretically, an increased private investment in agriculture cannot compensate for a decline in public investment in agriculture. In practice, there seems to be a fall even in the private investment in agriculture.

There is a decline even in non-farming sub-sectors of the agricultural sector in Andhra Pradesh. Livestock population has diminished over the last few years. Decline in the land area of pastures and destruction due to drought has created a scarcity of food and fodder for cattle and other livestock.

High debt liability is a serious challenge for many of the farmers in Warangal district. In this study, to understand the repayment capacity of the farmers, asset liability ratios have been calculated for each farmer in the sample considered. These ratios show that the asset values of most of the farmers are more than their liability values. But most assets are not liquid assets, and hence, hardly constitute the farmers' repaying capacity. In the case of the crops cultivated in the district, especially cotton and paddy, the material costs form a greater part of the cost than the labour costs. One of the reasons for high material costs is the expensive nature of chemical plant protection methods. The high-risk nature of farming has resulted in less sustainable short-run measures, which could prove to be very expensive from the long run perspective.

According to the data that I have analysed on the basis of information gathered from different sources like Agricultural institutions, Statistical departments of the state, as well as Warangal district, leads to the conclusion that agriculture in the district is in a way a losing proposition for it is not remunerative. The major crops of the District are Paddy, Jowar, Groundnut, Chillies etc. Though new and innovative schemes, policies and technologies have been introduced in the field of the Agriculture sector, adequate and expected goals of productivity levels are not being generated accordingly. Aspects like droughts, unseasonal rains, pest attack and fall in prices of the produced agricultural goods on the one hand and the mixed cropping pattern of the district with an overall cost of production which is higher than remuneration on the other, is creating losses in the Agriculture sector of the District. The average agricultural income of farmers in the district is not of the expected level because the cost of production is greater than the returns from the same.

In the prevailing era of high inflation, the cost of fertilizers, pesticides, seeds, and all other agricultural inputs have gone up to an extreme level, where a farmer cannot afford them even after nominal subsidies have been provided on them by the government. As most of the farmers in the District don't have an alternative source of livelihood, they still continue to depend on agriculture. Somehow, even if they manage to invest and cultivate, the returns for their production is not in the same proportion of their investments. Because as a whole, including the cost of both material and labour, the total cost of agricultural production has increased. The unremunerative prices have added to the situation. For the purpose of enhancing production and productivity and to fight against disease attacks on agricultural crops, farmers are forced to spend a great deal on manures, pesticides, fertilizers and chemicals. Added to this, the agricultural labour charges have also increased. The agricultural prices however, are not increasing in accordance with the cost of production. The removal of restrictions on imports in the globalization era in India, the low level of import tariffs levied on such goods and the removal of quantitative restrictions on such agricultural goods by the Indian Government has caused several agricultural imports to increase tremendously. This in turn has created a high competition for domestic agricultural products, with respect to quality and prices. This is a significant reason for the decline of minimum support prices to such domestic goods. Hence, the returns to production are not increasing in accordance with the cost of production. This is leading the farmers into debts and over dues and is resulting in farmer suicides.

Added to the above mentioned reasons, if we observe the situation of the agricultural district of Warangal in recent years, the cultivation of crops is mostly dependent upon natural sources of irrigation. The cultivation of crops in Warangal is thus based on the availability of adequate and timely rainfall. The acute and unprecedented drought and continuous decline in rainfall during the 1999 to 2004 period in the state, caused high distress levels among the district farmers due to a complete lack of sources of cultivation, as the farmers of the district depend mostly on monsoons. After experiencing such hard situations of drought, untimely showers and floods in the district yet again caused distress in the agrarian situation of the district. This has changed the fate of the once wealthy farmers of a state which was popularly known as the "Bowl of Rice" (Annapurna).

Although the government has announced loan waivers and reliefs, the distress caused in the district due to the above discussed reasons, is beyond repair. The bureaucratic delay in India always results in untimely help. Such delayed assistance is useless to the farmers because in times of need, their suffering is immense and the burden caused due to such situations forces some of the farmers to even take such extreme steps like killing themselves, thus throwing their families into the clutches of grief and troubles. In the recent decade , the agrarian district of Warangal has been termed as a "Hub of Farmer Suicides", as this recent period has displayed one of the highest records of farmer suicides in the country. During the last decade, a total number of 677 farmer suicides have been recorded by official statistics of the district. It is believed that there are a greater number of farmers who have committed suicide than have been recorded in the official list.

The announced debt reliefs, debt waivers and exgratia for the crop failures and farmer suicides etc., are still under the process of distribution even after a period of 4 years. This is therefore an imaginable situation with respect to the effectiveness of policies regarding the agriculture sector of the state and in turn of the Warangal district.

Suggestions

Ironic as it may sound, even in an era of unprecedented liberalization like the present one, an adequate response from the state seems to be the most crucial component of a set of possible remedial measures. However, in the context of this study, policy prescriptions are being made keeping in mind the fact that state failure has almost as much to do with the agrarian crisis, as market failure. Therefore, those measures are suggested that would help the state and the market complement and strengthen each other.

As much as the problems are multi faceted, the solutions that they demand have multiple aspects too. While technology can be of possible help, the role of the institutions cannot be underestimated. For any solution to be effective, the stakeholders have to be part of it, both at the decision – making level as well as at the implementation level. The formation of farmer co-operatives backed by local governing bodies and the adaptation of grass root level plan projects that can be implemented through these co-operatives form a basic step in addressing the problem. Mostly, priority has to be given to finding alternatives for irrigational sources which are highly required.

- *Strengthening of rural infrastructure:*

The withdrawal of the state is most visible in terms of the falling public investment in rural infrastructure. As stated earlier, a fall in public investment in agricultural infrastructure cannot be compensated by a corresponding rise in private investment, due to the complementary nature of public and private initiatives. These initiatives have a role to play in the strengthening of rural infrastructure and farmer co-operatives can, with the aid of local government bodies, function as the agencies that implement these projects. The advantages of decentralized planning and stakeholder involvement can be realized in such a scenario. The role of the state would be in providing credit guarantees and the necessary technical expertise in building and operating such projects. Apart from the long- run benefits of the infrastructure created, the short-run employment creation would also have a positive effect on the local economy.

- *Crop Insurance:*

The uncertain nature of agriculture makes crop insurance schemes imperative for any remedial package intended for the sector. The state may consider reducing the premium rates for small and marginal farmers at least during the initial years. The state should make sure that crop loss estimations are carried out in an efficient and scientific manner and that the payment of compensation should be within a stipulated and short time period. Farmer co-operatives should play a crucial role in raising awareness about such schemes among farmers.

- *Creating organized institutional credit facilities:*

The state can provide credit directly or through any of its agencies. The state may also simply provide guarantees to private sector credit providers. Credit required for meeting infrastructural requirements can be provided to the farmer co-operatives, thus doing away with the problem of lack of assets with individual farmers for providing collateral. Credit required for meeting the farmers' day to day expenses can be provided by the co-operatives on personal guarantees. Providing credit through the co-operatives would also keep a check on using borrowed funds for conspicuous consumption. Though there are co-operatives available, there is still a need for their effective functioning. Co-operatives can also acquire assets like tractors and members can use them for meeting their needs by paying only the operational costs.

- *Development of research and extension activities:*

Research and extension services should be undertaken in a more extensive way for the betterment of the agricultural sector in Andhra Pradesh. An institutional mechanism should be developed to make it a continuous process by which research is constantly oriented towards the needs of the farmer and the farmer is made fully aware of the various aspects of new technology. Agricultural officers of the state should play a crucial role in making sure that the developments on the technological front are utilized optimally by the farmers at the grassroots. Public investment in agricultural research and extension activities should increase sufficiently to meet this need.

- *Setting up of rural agro based industries:*

Setting up of rural agro based and other small scale industries will enhance alternate job opportunities in the rural areas. Inter-sectoral linkages that emerge from the setting up of such industries could be beneficial to the farming community as well. This could possibly remove those hindrances that negate the Lewisian presumption of the absorption of surplus agricultural labour. State agencies have a crucial role in providing education and training so as to facilitate inter-sectoral mobility of labour.

- *Development of agricultural information system:*

In a market-oriented system, what is of utmost importance is the provision of relevant and timely information to the farmer as an economic agent. The development of a reliable agricultural information system can help the farmers by providing necessary statistics, which will help them in making appropriate decisions. Key statistics on global, regional and local prices, data on cost and availability of inputs, information on annual and seasonal trends etc. will immensely benefit the farmers in being prepared to face the pressures of global market forces.

- *Encouragement of sustainable agrarian practices:*

Declining productivity of land in the post-green revolution era highlights the importance of following sustainable farming practices. The state should actively encourage the promotion of natural methods of agricultural protection including measures like integrated pest control management. Such efforts are considered useful in two ways. Firstly, it will reduce the input costs of the farmers. Secondly, resorting to natural methods of protection rather than to chemicals and other pesticides will reduce long-run damage to the soil and water bodies. This has a significant effect, especially in reducing the farmers' dependence on giant chemical corporations.

- *Impart training on quality improvement and maintenance:*

It is very important to ensure quality maintenance and improvement so as to have a good rate of agricultural exports amidst the cut throat competition of the liberalized market. Necessary training should be provided to the farmers with respect to various aspects of cultivation, storage and processing of the farm produce. The agencies of the state, the local government bodies and the farmer co-operatives all have a role to play in this regard.

The policy makers will have to look at various handicaps as challenges and challenges as opportunities. They will be successful in their tasks to the extent they have a clear idea about who their clients are, the time frame within which they are working and what the strength of the implementing mechanism is. Above all, they have to be cognizant of the political reality while contemplating any policy initiative. They will have to identify who gains and who will lose, their allies and their adversaries. Unfortunately, there are very few options in the policy arena that are without conflict. The policy makers have to be as sensitive to the political reality as they are to the economic and technical opportunities.

BIBLIOGRAPHY

Adelman, I (1986), "The Balance Between Industry and Agriculture in Economic Development" in proceedings of the International Economic Association, Delhi.

Bhattacharya, B, "Trade Liberalization and Agricultural Price Policy in India since Reforms" (Article), Indian Journal of Agricultural Economics, Vol. 58, No. 3, July-September, 2003.

Azadi, Muzaffar, "Globalization and the States: Interrogating the Farmers' Movement in India", Journal of Economic and Social Development, Vol. 4, Jan-Jun, 2002.

Madhu, Sangeeta, "Before the Final Step", The New Sunday Express, Sept 21, 2003.

Singh, Sukhpal, "Crisis and Diversification in Punjab Agriculture: Role of State and Agri Business", Economic and Political Weekly, Dec 25, 2004, Pp. 5583-5590.

Chowdhary, K.R. (2004), "Who's Life Was That Anyway…Farmers", "Suicides in Andhra Pradesh", Forum Against Globalization.

Chowdhary, K.R., Prasad Rao and M. Koteshwara Rao, "Distress of Farmers X-rayed: A Case of Andhra Pradesh", Andhra Pradesh Rythu Sangam, 2004.

Narayana, N.S. and Parikh, "Strategies for Agricultural Liberalisation: Consequences for Growth, Welfare and Distribution", Economic and Political Weekly, Sept 30, 1995, Pp. A90-A92.

Mario, Rutten and K.P. Kannan (1990), "Access of Big Farmers to Land and Agricultural Trade in Central Gujarat, India", in: Ghanshyam Shah (ed.), Capitalist Development; Critical Essays; Felicitation Volume in Honour of Prof. A.R. Desai, Bombay, Popular Prakashan; Pp. 230-252.

Misra and Puri (2000 ed.), "Indian Economy", Himalaya Publishing House, Mumbai.

Mohanty, B. B. and Sangeeta Shroff, "Farmers' Suicides in Maharashtra", Economic and Political Weekly, Dec 25, 2004, Pp. 5599-5606.

Halliburton, Murphy (1998), "Suicide: A Paradox of Development in Kerala", Economic and Political Weekly, Sept, 1998, Pp. 2341-2345.

Ashok Gulati and Shashanka Bhide, (1995), "What do Reformers Have for Agriculture", Economic and Political Weekly, May 6-13, 1995, Pp. 1089-1093.

Jha, Brajesh and Mohapatra, B.B., "Liberalization and Agricultural Prices: Some Disconcerting Trends" (Article), Indian Journal of Agricultural Economics, Vol. 58, No. 3, July-September, 2003, Pp. 375-386.

Dutt, Raddar and Sundaram K.P.M. (2001), "Indian Economy", S. Chand and Company Ltd., Ram Nagar, New Delhi.

Desai, Bhupat. M and M.V. Namboodiri, "Determinants of Total Factor Productivity in Indian Agriculture", Economic and Political Weekly, Vol. 32, No. 52, Dec 27, 1997, Pp. A165-A171.

Harris, Barbara White, "Agricultural Policy and its Implications for Agricultural Reforms", Journal of Economic and Social Development, Vol. 4, Jan-Jun 2002.

Lahiri, Sutapa, "Distress in Punjab Agriculture Under the WTO Regime: A Brief Note", Mainstream, Sept 1, 2001, Pp. 17.

Maris, Ronald. W, "Suicide" (Article), Encyclopedia of Sociology, 2nd Edition, Pp. 3077-2087.

Bailey, Bruce. E, "Suicide" (Article), International Encyclopedia of Sociology, Vol. 2, Pp. 1387-1390.

Chandrasekhar Rao. N, "Liberalization and Suicide of Farmers of India", Indian Journal of Agricultural Economics, Vol. 58, No. 3, July-September, 2003, Pp. 395-396.

Rao, V.M and D.V. Gopalappa, "Agricultural Growth and Farmer Distress: Tentative Perspectives from Karnataka", Economic and Political Weekly, Dec 25, 2004, Pp. 5591-5598.

Ramasamy, C, "Constraints to Growth in Indian Agriculture: Needed Technology, Resource Management and Trade Strategy", Indian Journal of Agricultural Economics, Vol. 59, No. 1, Jan-Mar, 2004.

Sen, Bhowani, "New Tasks in Agriculture", Mainstream, Vol. 40, Jul-Dec 2002.

Thirlwall, A.P., (1999, 6th Edition), "Growth and Development", Macmillan, London.

UNDP, "Human Development Report 2004", Oxford University Press, New Delhi.

"Human Development Report 2007, Andhra Pradesh", Vidya Graphics, Hyderabad.

The Indian Institute of Public Opinion, "Agriculture in the Reform Vortex", Monthly Commentary on Indian Economic Conditions, Vol. XLV, No. 7-8, Feb-Mar 2004, Pp. 535-536.

Woodward, Richard. T (2000), "Market-based Solutions to Environmental Problems", Journal of Agricultural Economics, Aug 2000, Pp. 259-266.

Rythu Sahaya Committee (1998), "Rythu Atmahatyalu", Report on Farmer Suicides in Warangal District (1997-98).

Sudershan Reddy, A and B Venkateshwar Rao (1998), "Gathering Agrarian Crisis – Farmers' Suicides in Warangal District, A Citizen's Report, Hanamkonda, A.P.

Revathi, E (1998), "Farmers' Suicides: Missing Issues", Economic and Political Weekly, Vol. XXXIII (20) May 16.

QUESTIONNAIRE

Research study on Agricultural Distress in Warangal District.

District:

Mandal:

Village:

I. General Information:

a) Name of the Farmer:
b) Family composition:
   Male: Female: Children: Total:
c) Social Status: SC/ST/BC/OC
d) Size Class of Farmer: Small/ Medium/ Large
e) Whether Tenant Farming: Yes/ No.

II. Asset:

1) Land:
a) Land (in acres):
   Owned Leased in: Leased out: Total:
   Irrigated:
   Dry:
   Total:

b) Land (in acres):
   Cultivable:
   Current Fallow:
   Permanent Fallow:

c) Land Value (in Rs./Acre)
   Irrigated:
   Dry:
   Total:

ii) Livestock:

	Number	Value
• Cow		
• Bullocks		
• Buffalo		
• Poultry		
• Sheep and Goats		
• Others		
Total:		

iii) Farm Machinery and Implements:

	Name	Value
•		
•		

iv) Source of Irrigation:

	In acres	Value
• Canal		
• Tank		
• Well		
• Bore well		
• Pump set		
• Others		

v) Buildings:

	Number	Value
•		
•		
Total:		

vi) Deposits:

- Bank
- LIC
- Others

vii)Liquid Amount:
- Cash in hand:

viii) Other Assets:

- Gold
- Silver
- T.V
- Radio
- Vehicles
- Gas stove
- Others

Total

Total Value of Assets:

III.Liabilities:

Loans:
- Sourcc:
- Amount:
- Purpose:
- Rate of interest:
- Amount Repaid:
- Amount Outstanding:

IV. Cropping pattern:

Crops cultivated Area in Acres

-
-
-
-

New Crops cultivated Area in Acres

-
-
-
-

V. Cost of Cultivation of Crops:

Material:

Cost per Unit

Total Cost

- Rent for the Land Lord
- Materials
- Seed
- Seedlings
- Organic manures
- Fertilizers
- Plant protection
- Chemicals
- Others

Total

Labour:

Cost per Unit

Total Cost

- Preparatory cultivation
- Digging of pits
- Sowing
- Planting
- Spreading out manures
- Application of Fertilizers

- Spraying
- Weeding
- Harvesting
- Threshing
- Processing
- Other operations

Total

VI. Returns

	Crop	Yield/Acre	Total Yield	Value
•				
•				
•				
Total				

Net Returns:

Crop
Gross returns
Cost of Cultivation
Net Returns
Loss

-
-
-
-

Total

VII. Opinion Survey:

- Whether your extent of land is sufficient for making out a decent living? Y/ N.
- Whether rainfall is sufficient for raising a crop? Y / N
- Whether irrigation facilities are available for your cultivation? Y / N.
- Whether quality inputs are available to you in the market? Y / N.
- Whether technical advice is given to you in the market? Y / N.
- Whether you face the problem of shortage of credit? Y / N.
- What are the problems in getting loans? Y / N.

- Whether banks are financing as much as you require for crop cultivation?Y/N.

a) Rate of interest
b) Crop of loans
c) Medium term and Long term Loans

- Whether you are borrowing from money lenders, traders etc? Y / N.
- What are the reasons for crop failure?

a) Shortage of rainfall/ drought
b) Lack of irrigation
c) Lack of quality inputs
d) Lack of credit
e) Lack of technical guidance
f) Due to pest and disease attack

- What are the marketing problems?

a) Is there demand for your products?
b) Where do you sell your products?
c) To whom?
d) What is the middlemen's share in consumers' price?
e) Consumer price per quintal?
f) How much farmer is getting?

- Are you getting remunerative price? Y / N
- How imports are affecting the demand for your products and price?